A FORMULARY OF PAINTS AND OTHER COATINGS

VOLUME II

Compiled by

Michael and Irene Ash

Chemical Publishing Co., Inc.
New York, N. Y.

Chemical Publishing Co., Inc.

ISBN 0-8206-0292-2

Printed in the United States of America

CONTRIBUTORS

American Cyanamid Co.	Wayne, New Jersey
Ashland Chemical Corp.	Minneapolis, Minnesota
Baker Castor Oil Co.	Bayonne, New Jersey
Day-Glo Color Corp.	Cleveland, Ohio
Eastman Chemical Products, Inc.	Kingsport, Tennessee
H.B. Fuller Co.	St. Paul, Minnesota
General Electric	Waterford, New York
Georgia Kaolin Co.	Elizabeth, New Jersey
ICI Americas	Wilmington, Delaware
Johns-Manville, Celite Div.	New York, New York
McCloskey Varnish Co.	Philadelphia, Pennsylvania
NL Industries	Hightstown, New Jersey
Polyvinyl Chemical Industries	Wilmington, Massachusetts
Rohm and Haas Co.	Philadelphia, Pennsylvania
A.E. Staley Manufacturing Co.	Decatur, Illinois

PREFACE

The physical aspects of paint chemistry and technology are well covered in many up-to-date texts and the reader is strongly advised to become acquainted with this material before using this formulary.

This volume and the ones that will follow are a catalogue of formulas reflecting the state of the art in the paint industry. It is meant to serve as a tool for anyone who must turn out quality paint products, without the fiscal luxury of developing completely new, untested prototypes. However, this is not a cookbook. Most of the formulas contained are starting-point preparations that do require refining. It is that giant leap from nothing to a workable idea that this series attempts to make.

Anyone familiar with coatings knows it is impossible to have distinct chapters dealing with only one product. Of the chapters presented here almost all contain major elements of the others. The descending logical sequence of each chapter is loosely built on:

A. Use (primer, enamel, exterior paint, interior paint, luminescent paint, etc.) and special characteristics
B. Vehicle (oil, alkyd, epoxy, latex, etc.)
C. Color (if any specified)

Unless otherwise specified, all formulas have the quantities of ingredients given in parts by weight. Where test results were available, they follow the formula. A list of abbreviations that are used throughout the formulary is included.

All constituents appearing by their trademark name are printed in boldface type, and the manufacturers' names and addresses appear after the list of alphabetized tradenames in the Appendix.

PREFACE TO VOLUME II

This second volume of A Formulary of Paints and Other Coatings has been prepared because of the popular response to the first volume. All the formulas in this volume are new and up-to-date. The original format of categorical breakdown has been adhered to as it provides the user with a simple means of locating specific formulas.

We gratefully acknowledge the contributors for their cooperation in providing the formulations that have made this book possible.

ABBREVIATIONS

a	approximately
b.p.	boiling point
C	degrees Celsius
cm^2	centimeters squared
cps	centipoise(s)
CPVC	critical pigment volume concentration
F	degrees Fahrenheit
ft^2	square feet
g	gram(s)
gal	gallon(s)
hr	hour(s)
in.	inch(es)
in-lb	inch-pounds
KU	Krebs Units
lb	pound(s)
max	maximum
min	minute
Min	minimum
no.	number
NV	nonvolatile
%	percent
P/B	pigment to binder
PBW	parts by weight
pH	hydrogen ion concentration
PHG	per hundred gallons
PVaC	polyvinyl acetate
PVC	pigment volume concentration
rpm	revolutions per minute
s	second(s)
sol'n.	solution
theor.	theoretical
visc.	viscosity
VNV	vehicle nonvolatile
vol.	volume
Wt.	weight

TABLE OF CONTENTS

Chapter I	Primers	1
Chapter II	Exterior Paints	58
Chapter III	Metal Paints	140
Chapter IV	Masonry Paints	156
Chapter V	Enamels	172
Chapter VI	Interior Paints	217
Chapter VII	Varnishes, Lacquers, and Wood Finishes	289
Chapter VIII	Specialty Coatings	303
Chapter IX	Industrial Paints	337
	Appendix	366
	Index	388

Chapter I

PRIMERS

A primer is the first "gluelike", pigmented coat that acts as the interface between the finished coat and the substrate. It must be wettable and penetrable to achieve substrate adhesion, while ensuring compatibility with the finish coats.

Interior Primer Sealer

Formula No. 1

(Acrylic)

	lb	gal
Water	393.0	47.18
Latekoll D (8%)	70.0	8.28
Daxad 30	5.0	0.52
Igepal CO-630	3.0	0.34
PMA-30	1.0	0.09
Titanox R-901	100.0	3.00
Veroc #1	85.0	3.77
Mica (325 mesh)	30.0	1.28

Add in order listed and mix on a high speed disperser for 10 min, then let down as follows:

Texanol	20.0	2.53
Ethylene Glycol	40.0	4.32
Acronal 290 D	249.6	28.69

PVC (%)	35.0
NV by Wt. (%)	35.1

NV by Vol. (%)	23.0
lb/gal	9.97
Visc. (KU)	74 ± 5

No. 2

(Latex)

	lb	gal
Charge the following to the tank:		
Tamol 731 (25%)	8.0	0.88
Triton CF-10	2.0	0.23
Colloid 600	1.0	0.13
Water	20.0	2.40
Hexylene Glycol	20.0	2.62
Ethylene Glycol	15.0	1.62
Super Ad-It	1.0	0.13
Hydroxyethyl Cellulose–2.5% (**QP-4400**)	30.0	3.60
Add the following at low speed, 1900 fpm:		
Ti Pure R-901	125.0	3.75
Snowflake	200.0	8.86
Grind on high-speed dispersion equipment for 10 min, then add at low speed:		
Water	20.0	2.40
Colloid 600	1.0	0.13
Rhoplex AC-22	440.0	49.72
Hydroxyethyl Cellulose–2.5% (**QP-4400**)	160.0	19.20
Water	50.0	6.00

Adjust pH to 9.5 with ammonium hydroxide (28%).

PVC (%)	39.2
Total Solids (%)	47.5
Visc.–Initial (KU)	86

Vinyl Acrylic Primer

Formula No. 1

(Sealer)

	lb	gal
KTPP	1	0.1
Water	100	12.0
Tamol 731-25	9	1.0
Triton X-102	4	0.5
Ethylene Glycol	20	2.1
Vicron 15-15	150	6.7
Cellosize QP-15,000 (2.5% sol'n.)	100	12.0
Ti Pure R-901	150	4.5
ASP-400	50	2.3
Mistron HGO-55	125	5.4
Metasol 57 (sol'n.)	2	0.3
Foamicide 677	2	0.3
Cellosize QP-15,000 (2.5% sol'n.)	100	12.0
Texanol	16	2.0
Varaqua 936	318	35.0
Water	75	9.0

PVC (%)	49.8
Total NV by Wt. (%)	54.1
lb/gal	11.62
Visc. (KU)	95-100

No. 2

(Blister and Stain-Resistant)

Water	175.0
Daxad 30	10.0
Nopco NDW	1.0
Metasol 57	1.8
Lead Silicate	115.0
Zopaque R-88	125.0
Silver Bond B	100.0

Mix on a high speed mill and add:

Texanol	7.0
Ethylene Glycol	35.0
Natrosol 250 MR	5.0
Nopco NDW	1.0
Par Cryl 400	495.0
Water	45.0
Ammonium Hydroxide	2.0

Yield (gal)	100
PVC (%)	30
Solids (%)	52
Visc. (KU)	74-80
pH	9.5

No. 3

(Blister and Stain-Resistant)

Water	205.0
Daxad 30	10.0
Nopco NDW	1.0
Metasol 57	1.8
Lead Silicate	115.0
Titanox RA-NC	125.0
Silver Bond B	100.0

Mix on high speed mixer and add:

Texanol	7.0
Ethylene Glycol	35.0
Natrosol 250 MR	5.0
T&W 345-100 Super Alkyd	25.0
Lead (24%)	1.0
Manganese (6%)	0.5
Nopco NDW	1.0
ParCryl 400	440.0
Water	40.0
Ammonium Hydroxide	2.0

Yield (gal)	100
PVC (%)	30
Solids (%)	52
Visc. (KU)	74-80
pH	9.5

Interior Wall Primer

(Latex–Fast-Drying)

	lb	gal
Charge into mixer under agitation:		
Water	100.0	12.0
Ethylene Glycol	10.0	1.0
KTPP	1.2	0.1
Colloid 677	2.5	0.3
Strodex SEK-50	6.0	0.7
ASP-400	120.0	5.7
Cellosize QP-4400 (1½% sol'n.)	132.0	15.7
Titanox 2032	75.0	2.2
Atomite	150.0	6.7
Cosan 635W	1.0	–
Disperse in high speed mill then add slowly with agitation:		
Water	117.0	14.0
Colloid 677	1.5	0.2
Ucar Latex 365 or **366**	384.0	42.7

Note:

Protect thickener solution with 0.08% **PMA-30** if stored prior to use.

VPC (%)	40
CPVC (%)	52
Visc. (KU)	75-80
lb/gal	10.8
Total Solids (%)	45.5

Alkyd Spray Primer

(Fast-Dry Enamel–Sand Grind)

	lb	gal
Varcopol 470-60	114	15.0
VM&P Naphtha	195	31.0
M-P-A 60	4	0.5
Desertalc 55	150	6.5
ASP-400	175	8.1
Titanox RA-50	150	4.3
Xylol	43	6.0
Sand grind		
Varcopol 470-60	228.0	30.0
Cobalt Naphthenate (6%)	1.4	0.2
Exkin #2	0.6	0.1

PVC (%)	43.2
Vehicle NV by Wt. (%)	35.5
Total NV by Wt. (%)	64.2
lb/gal	10.43
Visc. (KU)	60–65

Anticorrosive Styrene-Acrylic Primer

Formula No. 1

(Spray)

Disperse	
Zinc Chromate ZCYT 746	42.0
Red Oxide 15866	41.0
Bermocoll E320GK (4% aq. sol'n.)	29.0
Calgon S (10% aq. sol'n.)	11.0
Then add, in order	
Ammonia (0.880–0.885 s.g.)	11.0
Water	11.6
Vinacryl 7175	853.0
Hercules Defoamer 1512M	1.0
Acticide MPM	0.4

PVC (%)	5.5
Total NV by Wt. (%)	51
Visc. at 25 C (cps)	60

No. 2

(Brushing)

Mix

Bermocoll E320GK (4% aq. sol'n.)		48.0
Calgon S (25% aq. sol'n.)		28.0
Hercules Defoamer 1512M		0.1
Mix, then add to the above		
Vinacryl 7175		522.0
Potassium Hydroxide (50% aq. sol'n.)		10.0
Then add quickly, with minimum shear		
Zinc Chromate ZCYT 746		43.0
Red Oxide 15886		80.0
Talc AT1		268.4
Hercules Defoamer 1512M		0.1
Acticide MPM		0.4
PVC (%)	34.5	
Total NV by Wt. (%)	66.6	
Visc. at 25 C (cps)	≈ 460	

No. 3

(White–Brushing or Spray)

Disperse

Bermocoll E320GK (4% aq. sol'n.)		66.0
Calgon S (10% aq. sol'n.)		18.0
Water		80.0
Zinc Phosphate		158.0
Tioxide R-CR2		103.0
Then add, in order		
Ammonia (0.880–0.885 s.g.)		10.5
Vinacryl 7175		564.0
Hercules Defoamer 1512M		0.1
Acticide MPM		0.4
PVC (%)	21.9	
Total NV by Wt. (%)	54.6	
Visc. at 25 C (cps)	≈ 420	

Acrylic Corrosion-Inhibiting Primer

	Formula No. 1		No. 2		No. 3	
	(Lead Silico Chromate)		*(Calcium Zinc Molybdate)*		*(Strontium Chromate)*	
	lb	gal	lb	gal	lb	gal
Grind paste:						
Neocryl A-601	383.1	45.59	—	—	—	—
Dispersant	5.8	0.70	—	—	—	—
			2.5	0.26	2.5	0.26
Antifoam	3.1	0.43				
			2.5	0.34	2.5	0.34
Red Iron Oxide	29.3	0.93	—	—	—	—
Lead Silico Chromate	263.8	7.74	—	—	—	—
Mica (325 mesh)	58.7	2.55	—	—	—	—
Titanium Dioxide	—	—	178.9	5.37	178.9	5.37
Calcium Zinc Molybdate	—	—	57.9	2.32	—	—
Strontium Yellow	—	—	—	—	71.8	2.32
Water	—	—	222.9	26.79	222.9	26.79
Propylene Glycol	—	—	31.0	3.59	31.0	3.59

Procedure:

Grind on high speed disperser 10–15 min/3600–4500 fpm

Let down:

Neocryl A-601	203.4	24.20	483.7	57.56	483.7	57.56
Propylene Glycol	27.0	3.13	–	–	–	–
Antifoam	2.7	0.37	–	–	–	–
Ammonium Hydroxide (28%)	2.3	0.31	–	–	–	–
Neutralizer	–	–	24.8	3.18	24.8	3.18
Thickener	–	–	5.0	0.59	5.0	0.59
	116.8	14.05	–	–	–	–

Total NV by Wt. (%)	50.0	41.6	42.1
PVC (%)	35	31	31
P/B Ratio	65/35	60/40	62/38
lb/gal	11.0	10.1	10.2
Visc. (cps)	550	500	500
pH	9.0	9.5	9.5

No. 4

(Lead Silico Chromate)

	lb	gal
Grind paste:		
Water	19.2	2.31
NeoCryl A-620	383.1	45.05
Dispersant	5.8	0.70
Antifoam	3.1	0.43
Red Iron Oxide	29.3	0.93
Lead Silico Chromate	263.8	7.74
Mica (325 mesh)	58.7	2.55

Grind on high speed disperser 10-15 min/3600-4500 fpm.
Let down:

	lb	gal
Water	10.2	1.23
NeoCryl A-620	203.4	23.92
Propylene Glycol	27.0	3.13
Antifoam	2.7	0.37
Ammonium Hydroxide	2.3	0.31
Thickener	116.8	14.50

Total NV by Wt. (%)	52.0
PVC (%)	32
P/B Ratio	61/39
lb/gal	11.0
Visc. (cps)	4100
pH	9.1

Note:

The addition of **Drew XPD** 11-0432 59.0 (10% NV aqueous slurry) eliminates flash rusting. The formulation is then called 044D.

No. 5

	lb	gal
NeoCryl A-621	298.95	35.25
Defoamer	2.29	0.31
Dispersant	5.68	0.52

Lead Silico Chromate	141.24	4.15
Red Iron Oxide	141.24	3.30
Barytes #1	70.66	1.93
Thickener	5.67	0.28
Surfactant	1.56	0.17

Disperse with high speed impeller (10 min @ 2000 fpm).
Premix and add the following:

NeoCryl A-621	347.83	41.02
Defoamer	1.10	0.14
Surfactant	1.28	0.17
Flow Agent	1.55	0.20

After uniform mixture is attained, add:

Nytal 300	298.21	12.56

Reduce this formula 4 to 1 with water to attain spray viscosity (25 s–#4 Ford Cup).

Total NV by Wt. (%)	58
PVC (%)	36.6
P/B Ratio	74/26
Fineness of Grind	6 NS
Visc. (cps @ 25 C)	57,000
pH	8.8
lb/gal	13.17

Alkyd Corrosion-Inhibiting Primer

316 Micaceous Iron Oxide	220.0
1428 Zinc Chromate	80.0
H.O. Zinc Oxide	80.0
1681 Red Oxide	220.0
C3000 Micro Mica	235.0
P-296-60 Soya Alkyd (TT-R-266, Type I, Class B)	318.0
Mineral Spirits	157.0
Bentone 38 (15%)	63.0
Troykyd (6% naphthenate)	2.1

Troykyd (24% lead naphthenate)		5.3
Troykyd (5% calcium naphthenate)		2.5
Troykyd (antiskin)		4.5
Solids by Vol. (%)	52.3	
PVC (%)	49.6	
lb/gal	13.9	
Visc. (KU)	83	

Chlorinated Rubber Primer

Formula No. 1

Oncor M-50	850.0
C.P. Chromium Oxide Green	93.6
Phthalocyanine Green	6.7
M-P-A 60 (mineral spirits)	15.3
Dyphos	8.5
Copolymer 186	291.0
Parlon S-20 or **Alloprene 20** (30% NV in 80% **Solvesso 100**/ 20% **Cellosolve**)	607.0
Epichlorohydrin	1.0
Solvesso 150	178.0*
Lead Naphthenate (24%)	3.6**
Cobalt Naphthenate (6%)	1.5**
Manganese Naphthenate (6%)	0.7**
Antiskinning Agent	3.0

* For faster dry and/or spraying replace with **Solvesso 100.**

** Use octoate in place of naphthenate driers for better solubility and stability in chlorinated rubber.

Weight Ratio (Pigment/NV Vehicles)	67/33
Visc. (KU)	83
PVC (%)	≈ 32

Very fast dry coupled with good flexibility and excellent chemical resistance; grind at 130 F.

The following pigment modifications are suggested to obtain alternate colors to the dark green submitted top coats.

Maroon: Reduce **Oncor M-50** to 650 parts and replace the chromium oxide and phthalocyanine greens by 300 parts of iron oxide.

Gray: Reduce **Oncor M-50** to 605 parts and replace the chromium oxide and phthalocyanine green by 145 parts **Titanox Ranc**, 48 parts zinc oxide, 95 parts magnesium silicate, 8 parts lampblack, and 4 parts phthalocyanine blue.

	No. 2	No. 3
Alloprene 20	13.5	10.0
Paralac IOW	9.0	–
Boiled Linseed Oil	4.0	–
Dehydrated Castor Oil (40 poises)	–	15.0
Red Lead or Calcium Plumbate	22.7	25.0
Synthetic Iron Oxide	–	2.0
Rutile Titanium Dioxide	–	2.0
Winnofil PI	7.2	5.0
Aromasol H*	35.4	29.8
Mineral Spirits	7.7	9.9
Thixatrol ST	0.5	0.5
Lead Naphthenate	–	0.6
Cobalt Naphthenate	–	0.2

Note:
Should be used away from areas of cathodic protection.

* A blend of **Solvesso 100** and **Solvesso 150** may be substituted for the **Aromasol H.**

No. 4

Mill base grind (3 roll mill, 110 F):

Parlon S-20	10.20
Red Lead (97%)	18.60
Zinc Chromate	3.10
Mica	3.10
Red Iron Oxide	0.50

Talc	7.00
Xylene	20.00
M-P-A 60 (xylene)	0.00 (A) and 16.50 (B)

Let down:

Aroclor 1254	5.60
Aroclor 5460	2.70
Xylene	27.90
Epichlorohydrin	0.05

No. 5

	lb	gal
Alloprene X-20	112.3	8.32
Xylene	169.1	23.48
Bentone 38	13.6	0.96
Xylene	112.7	15.65
Cerechlor 42-P	84.6	10.57
Syloid ZN-1	14.0	0.84
Federated # 111	1349.3	22.87
Ferrophos HRS2131	900.3	16.52
Y539D	23.1	0.79

Procedure:

Dissolve resin in solvent. Pregel **Bentone 38** and portion of xylene on Cowles disperser. Add pregelled **Bentone 38** to solvent/resin blend and mix at medium speed for 10 min. Add **Cerechlor** and **Syloid** and grind at medium speed for approximately 10 min. Add **Y539D**, **Ferrophos**, zinc dust and grind at high speed approximately 20 min. Maintain cooling bath to avoid overheating.

No. 6

Parlon S-20	2.48
Aroclor 5460	0.67
Aroclor 1254	1.07
Thixatrol ST	0.15
Dioctyl Sebacate	0.26

Xylene	4.08
Super High-Flash Naphtha	2.59
Turpentine	0.70
Federated No. 555 Zinc Dust	88.00

lb/gal	34.67
Total Solids (% by Wt.)	92.63
Total Solids (% by Vol.)	64.76

It is recommended that the primer be topcoated for optimum performance.

Procedure:

Dissolve the resins and plasticizers in about one-half of the solvents. Mill in the **Thixatrol ST** so that a temperature of about 120-130 F is obtained and then add remaining solvent.

The finished paint is prepared by slowly adding the zinc dust to the base vehicle with good agitation.

No. 7

	lb/100 gal
Parlon 520	145.00
Koppers Rezyl 869	72.00
Aroclor 6062	72.00
Dyphos	5.75
316 Micaceous Iron Oxide	520.00
Lampblack	1.90
Titanium Dioxide	9.50
Bentone #38 Mastergel (15%)	30.00
Solvesso 100	343.00
Mineral Spirits	86.00
Epichlorhydrin	0.75
316 Micaceous Iron Oxide	5.20

Procedure:

The lampblack and titanium dioxide is pebble milled in portion of vehicle. The micaceous iron oxide is stirred in (no grinding).

4-mil (dry) film coat over 1-mil (dry) anticorrosive primer suggested.

Total Solids by Vol. (%)	37
PVC (%)	32
lb/gal	12
Visc. (KU)	75
Pigment/Vehicle Ratio (%)	44.6-55.4
	35.7

Epoxy Gray Baking Primer

Rutile Titanium Dioxide	89.5
Lampblack	2.0
Urea-Formaldehyde Resin (60% NV)	183.0
Cellosolve Acetate	87.9
Toluene	87.4
Araldite 597 ET-55	466.6

Alkyd Zinc Chromate Primer

(Fast-Dry, Nonlifting)

	lb	gal
Y-539D	200.0	4.40
Kadox 515	30.0	0.53
Desert Talc 57	100.0	4.32
Celite 281	50.0	2.58
Bentone 38	3.0	0.30
Ethyl Alcohol	3.0	0.50
Lead Naphthenate (24%)	4.0	0.50
Varkyd 340-50X	300.0	37.00
Xylene	100.0	14.00

Grind the above, then add:

Varkyd 340-50X	138.0	17.00
Xylene	50.0	7.00
Cobalt Naphthenate (6%)	4.0	0.45
Exkin #2	0.5	–

Above paste to be thinned for brushing use:

Pentoxone	52.0	7.50
Visc. (KU)	82	

For spraying use:

Xylene to reduce to spray viscosity.

PVC (%)	33.6
lb/gal (paste)	11.1

Epoxy Zinc-Rich Primer

Formula No. 1

	lb	gal
A **Resypox 1571**	112.4	11.18
MIBK	67.3	10.03
Cellosolve	67.3	8.67
Xylene	67.3	9.32
Cab-O-Sil	25.4	1.45
Butle 216-8	9.4	1.11
B **Resymide 1115**	50.9	6.13
MIBK	28.2	3.63
Xylene	28.2	3.90
Zinc Dust	2372.0	40.32

lb/gal	28.6
PVC (%)	70.0
Total Solids (%)	89.8

No. 2

(Dust)

	lb/100 gal
Araldite 488 N-40	408.0
Zinc Dust 430	1323.6
M-P-A 60 (xylene)	13.2

Xylene		86.0
Methyl Isobutyl Ketone		43.0
Cellosolve		43.0
PVC (%)	60.5	
Zinc Dust (%)	90.0	
Total NV by Wt. (%)	77.7	
Total NV by Vol. (%)	38.1	

Procedure:

Add solvents to the 488 resin solution. Then add the zinc dust pigment with high speed agitation.

Visc.–Initial (KU)	100

No. 3

	lb	gal
RCI 93-905	149.0	18.40
Solvent Blend*	338.1	48.30
Cabosil M5	12.1	0.67
Aerosil R972	23.9	1.33
Calcium Oxide	7.2	0.30
Ferrophos HRS2131	555.9	10.20
Federated #111	826.0	14.00
Solvent Blend*	41.3	5.90
Cobalt Naphthenate (6%)	2.7	0.35
Manganese Naphthenate (6%)	2.7	0.35
Exkin #2	1.5	0.20

* Solvent Blend

VM&P Naphtha-Rule 66	50
Cellosolve Acetate	25
Butyl **Cellosolve**	15
Toluene	10

Procedure:

Mix resin, **Cabosil** and **Aerosil** at medium speed on Cowles till evenly dispersed. Add calcium oxide, **Ferrophos** and zinc dust and grind at high speed for approximately 30 min. Keep temperature under 120 F with cooling jacket, if necessary. Premix driers, antiskinning agent, and solvent and add as let down. Mix for 3 min at low speed.

No. 4

(Dust)

	lb	gal
RCI 93-905	167.2	20.64
Cabosil M5	12.1	0.67
Aerosil R972	25.0	1.39
Calcium Oxide	8.9	0.37
Federated #111	558.7	9.47
Ferrophos HRS2131	836.6	15.35
Y539D	11.7	0.40
Solvent Blend*	355.6	50.80
Cobalt Naphthenate (6%)	2.7	0.35
Manganese Naphthenate (6%)	2.7	0.35
Exkin #2	1.6	0.21

* Solvent Blend

VM&P Naphtha Rule 66	50
Cellosolve Acetate	25
Butyl Cellosolve	15
Toluene	10

Procedure:

Into **RCI 93-905**, add **Cabosil M5**, **Aerosil R972** and mix at medium speed on Cowles disperser for 10 min (use solvent blend as needed). To this mix, add calcium oxide, zinc dust, **Ferrophos**, and **Y539D** and grind at high speed for 15 min (use solvent blend as needed). Premix 6% Co, 6% Mn, and **Exkin** and add to any remaining solvent. Add to above mix at low speed for 1 min.

Epoxy-Polyamide Zinc Dust Primer

		lb	gal
A	**Epon 1001**	93.1	9.26
	Solvent Blend*	72.2	10.03
	Bentone 27	20.0	1.27
	Solvent Blend*	70.7	9.82
	Ethanol	10.0	1.50
	Beetle 216-8	8.3	0.98

	Syloid ZN-1	16.8	0.96
	Y539D	20.0	0.69
	Ferrophos HRS2131	808.2	14.83
	Federated #111	1197.7	20.30
	Solvent Blend*	124.6	17.30
B	**Versamid 115**	49.7	6.16
	Solvent Blend*	49.7	6.90

* Solvent Blend

MIBK	33-1/3
Cellosolve	33-1/3
Xylene	33-1/3

Procedure:

Part A—Dissolve resin in solvent blend. Pregel the **Bentone** in the solvent blend and ethanol, then add to the resin solution, mix at low speed on Cowles till homogeneous. Add **Beetle** resin and mix at low speed for 2 min. Add **Syloid** and zinc yellow and mix at medium speed for approximately 5 min. Add **Ferrophos**, zinc dust, and enough solvent blend to obtain a good grinding viscosity. Grind for 30 min. Keep cooling jacket on to avoid overheating. Add remaining solvent, if any.

Part B—Blend **Versamid** and solvent blend and add to Part A base paint. Filter and wait 30-60 min before use. Potlife is at least one 8-h day.

Latex Red Iron Oxide, Zinc Chromate Primer

	lb	gal
KTPP	2.0	0.1
Ethylene Glycol	75.0	8.0
Tamol 850	11.0	1.3
Defoamer	3.0	0.4
Water	235.0	28.2
Red Iron Oxide	220.0	5.1
Zinc Chromate	45.0	1.5
Clay	50.0	2.2
Talc	120.0	5.2
Natrosol 250 MR	2.5	0.2
Biocide	1.0	0.1

Defoamer	1.0	0.1
Carbitol	10.0	1.2
Varaqua 939	425.0	47.2
Water	25.0	3.0

PVC (%)	37
Vehicle NV (%)	30
Total NV by Wt. (%)	59
lb/gal	11.7
Visc. (KU)	70–80

Latex Zinc Oxide Primer

(Alkyd Modified)

	lb	gal
Water	53.0	6.36
Thickener (3%)	102.0	12.30
Ammonium Hydroxide	1.0	0.10
Strodex MOK-70	4.0	0.41
KTPP	4.0	0.45
Nonmercurial Bactericide	2.0	0.16
Colloid 677	1.0	0.13
Polyglycol P-1200	2.0	0.24
Zinc Oxide AZO-33	100.0	2.14
Rutile Titanium Dioxide	185.0	5.60
Mica (325 W/G)	25.0	1.08
$CaCO_3$	35.0	1.51
Talc	35.0	1.51
Ethylene Glycol	26.0	2.80
Carbitol	12.0	1.28
1271 Alkyd (Ashland 100% solids)	50.0	6.00
Water	52.0	5.72
Igepal CO-710	3.0	0.34
Zirconium Drier (6%)	4.0	0.54
Manganese Drier (6%)	0.5	0.06
Cobalt Drier (6%)	0.5	0.06
Water	112.0	13.44

Colloid 677	1.0	0.13
Latex Reichhold 40-124	335.0	37.65

lb/gal	11.6
PVC (%)	30.45
Visc.–Initial (KU)	77

Phenoxy Zinc Primer

Formula No. 1

(Dust)

	lb	gal
PKHH	160.4	16.37
Solvent Blend*	380.5	54.05
MPA-60 T	16.0	2.20
Syloid ZN-1	9.4	0.56
Federated #111	906.8	15.37
Ferrophos HRS2131	605.5	11.11
Y539D	9.9	0.34

Procedure:

Dissolve resin into solvents. Add **MPA-60T** and **Syloid ZN-1** and mix on Cowles at medium speed for 10 min. Add **Ferrophos**, zinc dust, and zinc chromate and grind at high speed for 25 min. During grind obtain maximum 125 F.

* Solvent Blend

Methyl Ethyl Ketone	45
Toluene	45
Cellosolve Acetate	10

No. 2

Bakelite PKHH	10
Zinc Dust	90
MEK	18
Toluene	18
Cellosolve Acetate	4

Note:

Thin to spray viscosity with a 45/45/10 weight ratio blend of MEK/toluene/**Cellosolve** acetate.

PVC Zinc Dust Primer

Formula No. 1

	lb	gal
Solvesso 100	131.0	18.00
Toluene	132.0	18.26
Vinoflex MP 400 (medium)	77.0	7.45
Zinc Dust	1500.0	25.50
Aerosil R-972	35.0	1.83
Solvesso 100	100.0	13.73
VM&P Naphtha	99.0	15.76

Procedure:

Add in order listed and disperse on high speed disperser.

PVC (%)	78.6
Total NV (%)	77.7
Total NV by Vol. (%)	34.6
lb/gal	20.6
Visc. (KU)	75
Zinc/Binder Ratio	95/5

No. 2

	lb	gal
Solvesso 100	100.0	13.73
Toluene	251.0	34.72
Vinoflex MP 44 (medium)	176.0	17.04
Zinc Dust	1000.0	17.00
Aerosil R-972	35.0	1.82
Solvesso 100	100.0	13.73
VM&P Naphtha	144.0	19.15

Procedure:
Add in order listed and disperse on high speed disperser.

PVC (%)	52.5
Total NV (%)	67.1
Total NV by Vol. (%)	30.6
lb/gal	16.23
Visc. (KU)	100
Zinc/Binder Ratio	85/15

PVC Zinc Chromate Primer

Formula No. 1

	lb	gal
Xylene	250.0	34.48
Vinoflex MP 400 (medium)	150.0	14.52
Acronal 4F (100%)	50.0	5.71
Zinc Yellow 45-228	250.0	8.57
Nytal 300	30.0	1.26
Titanox CLNC	40.0	1.17
Cellosolve Acetate	50.0	6.17
Mineral Spirits	125.0	19.58
Bentone 38	3.0	0.20
Methyl Alcohol (95%)	2.0	0.29

Procedure:
Grind 16 h in a high density alumina jar mill.

PVC (%)	35.6
Total NV (%)	55.1
Total NV by Vol. (%)	34.2
lb/gal	10.3
Visc. (KU)	95
Reduced Visc., s DIN #4 (40% reduction by volume with toluene)	33

No. 2

	lb	gal
Xylene	250.0	34.48
Solvesso 100	100.0	13.73
Mineral Spirits	250.0	39.43
Santicizer 8	18.0	1.81
Vinoflex MP 400 (medium)	180.0	17.42
MPA 60 (xylene)	20.0	2.75
Oncor M-50	200.0	5.86
Atomite	50.0	2.21
Barytes X-5-R	75.0	2.10
Titanox CL	25.0	0.73

PVC (%)	34.2
Total NV by Wt. (%)	47.9
Total NV by Vol. (%)	26.4
lb/gal	9.69
Visc. (KU)	102*
Reduced Visc., s DIN #4 Cup (20% reduction by volume with xylene)	40

---

* May be let-down with xylene for satisfactory spraying viscosity.

Silicate Zinc Dust Primer

Formula No. 1

	lb	gal
Silbond H12A	534.3	63.23
325WG Mica	48.3	2.10
Bentone 38	30.8	2.17
Antiterra U	16.7	2.14
Federated #111	687.9	11.66
Ferrophos HRS2131	458.9	8.42
Y539D	5.8	0.20
Xylene	72.6	10.08

Procedure:

Into approximately 100 parts H12, add mica, **Bentone**, and **Antiterra U** and mix at low speed for 5 min on the Cowles. Add zinc chromate, **Ferrophos**, and zinc dust to above, grinding approximately 20 min at high speed. Viscosity may be adjusted during the grind by using remaining portion of H12 and/or xylene. Keep mix cool during grinding operation. Add remaining resin and/or solvent and mix at low speed for approximately 3 min.

No. 2

		lb	gal
A	**Silbond H6**	503.2	60.62
	Celite 499	44.2	2.30
	Bentone Pregel*	68.3	8.98
B	**Y539D** Zinc Yellow	30.4	1.04
	Ferrophos HRS2131	625.1	11.47
	Federated #111	919.8	15.59

* Bentone Pregel

Xylene	285
Bentone 27	30
Ethanol/Water (95/5)	12

Procedure:

Add pregel and **Celite** to approximately half of the silicate resin. Mix for 5 min at medium speed on the Cowles. Premix zinc dust, **Ferrophos**, and zinc yellow. Add this composite to the silicate, **Bentone-Celite** solution. Mix at medium speed for 5-10 min. Add remaining resin at low speed to above mix. If necessary for spray, use **Cellosolve** solvent.

No. 3

		lb	gal
A	**Silbond H6**	506.3	61.00
	Celite 499	44.2	2.30
	Bentone Pregel*	64.6	8.50

B	**Y539D** Zinc Yellow	46.4	1.60
	Ferrophos HRS2131	861.0	15.80
	Federated #111	626.0	10.80

* Bentone Pregel

Xylene	285
Betone 27	30
Ethanol/Water (95/5)	12

Procedure:

Add pregel and **Celite** to approximately half of the silicate resin. Mix for 5 min at medium speed on the Cowles. Premix zinc dust, **Ferrophos**, and zinc yellow. Add this composite to the silicate, **Bentone–Celite** solution. Mix at medium speed for 5-10 min. Add remaining resin at low speed to above mix. If necessary for spray, use **Cellosolve** solvent.

No. 4

Sodium Silicate	8.5
Water	2.3
Bentonite	0.5
Pigments	
Zinc Dust	6.3
Red Lead	1.8

Procedure:

The components can be mixed on the jobsite, strained through a cheesecloth and sprayed on the sandblasted steel to a dry film thickness of 3 mils. After an initial half hour drying period, the coating can be cured by spraying with a solution of:

Isopropyl Alcohol	75%
Phosphoric Acid (85%)	25%

After 24 h the coating is very hard and water insoluble.

Phenolic Resin Primer

Formula No. 1

(Intermediate)

	lb	gal
Basic Pb Silico Chromate	361.67	10.60
Barium Sulfate	62.00	1.69
M-P-A-60 (xylene)	0.31	0.04
Bakelite VAGD	106.47	9.35
Bakelite VYHD	32.88	2.90
Flexol 10-10	24.11	3.00
MEK	313.14	46.67
Cyclohexanone	100.20	12.70
Xylene	93.94	13.05

Procedure:

Dissolve vinyl resin in solvents, add pigments and grind on Cowles to Hegmen 5.

PVC (%)	44.61
Total NV (%)	53.65
lb/gal	10.0
Total Solids by Vol.	27.6
Total Pig. by Wt. (%)	38.7
Visc. (KU)	80.0

No. 2

(Red Lead)

	lb	gal
Read Lead (98% Pb_3O_4)	247.0	3.29
M-P-A 60 (xylene)	8.2	1.13
VAGH	162.8	14.07
TCP	16.8	1.73
MBK	273.2	40.47

MEK	105.5	15.72
Toluene*	75.1	10.43
VM&P Naphtha	81.1	13.19

* Technical grade toluene must be free of other contaminants such as water.

Procedure:

Dissolve VAGH Resin in solvents. Add pigment and grind in pebble mill overnight. Discharge and allow 72 h for viscosity to level.

Note:

VAGH-2 and VAGD are not acceptable for this specification. VAGH in the viscosity range of 250 to 400 cps (solution test UCC) is required.

lb/gal	9.7
Total NV (%)	44.22
Total Solids by Vol. (%)	19.34
PVC (%)	17.0
Visc. (KU)	75

Triamino-Ester Primer

Formula No. 1

(General Purpose, Red Lead)

	lb	gal
Red Lead (97%)	750.0	10.20
#680 Red Iron Oxide	175.0	4.90
Nytal 300 Talc	160.0	6.74
Asbestine 3X Talc	60.0	2.53
Bentone 38 Gel (10% in mineral spirits)	20.0	2.85
Chemacoil TA-100 (100%)	167.7	21.50
Chemacoil TA-303 (80%)	209.8	27.50
Mineral Spirits	164.1	25.25
Cobalt Drier (6%)	2.0	0.25
Lead Drier (24%)	9.7	1.00
Exkin #2	1.0	0.13

Visc.—Fresh (KU)	98
PVC (%)	36.7
Total Solids by Wt. (%)	86.3
Pigment Solids by Wt. (%)	66.7
lb/gal	16.73

No. 2

(Fast-Dry, Red Lead)

	lb	gal
Red Lead (97%)	1275.0	17.34
Bentone 38 Gel (10% in mineral spirits)	20.0	2.85
Chemacoil TA-303 (80%)	316.6	41.50
Mineral Spirits	251.9	38.75
Cobalt Drier (6%)	2.0	0.25
Lead Drier (24%)	4.8	0.50
Exkin #2	1.0	0.13

Visc.—Fresh (KU)	77
PVC (%)	35.7
Total Solids by Wt. (%)	81.6
Pigment Solids by Wt. (%)	68.1
lb/gal	18.47

No. 3

(General Purpose, Zinc Chromate)

	lb	gal
X-2127 Zinc Chromate Pigment	275.0	9.60
YLO-3288 Yellow Iron Oxide	130.0	3.87
414 Zinc Oxide	120.0	2.56
Nytal 300 Talc	175.0	7.37
Bentone 38 Gel (10% in mineral spirits)	20.0	2.85
Chemacoil TA-100 (100%)	163.8	21.00
Chemacoil TA-303 (80%)	206.0	27.00
Mineral Spirits	175.5	27.00

Cobalt Drier (6%)	2.0	0.25
Lead Drier (24%)	7.2	0.75
Exkin #2	1.0	0.13

Visc.–Initial (KU)	92
PVC (%)	36.0
Total Solids (%)	80.6
Pigment Solids (%)	54.6
lb/gal	12.47

Acrylic Industrial Hardboard Primer

	lb	gal
Titanium Dioxide	50.0	1.50
Carbon Black Dispersion	1.0	0.10
Clay	250.0	11.65
Propylene Glycol	50.0	5.80
Water	150.0	18.03
Dispersant	2.0	0.24
Defoamer	2.0	0.30

Grind 10 min at 4500 fpm to Hegman #7.

NeoCryl A-620	447.3	52.62
Defoamer	2.0	0.30
Water	111.8	13.44

(adjust with ammonium hydroxide to pH 9.8).

Total NV by Wt. (%)	45
Total NV by Vol. (%)	34
PVC (%)	40
P/B ratio	65/35
lb/gal	10.2
Visc. (s–#4 Ford Cup)	22
pH	9.8

Acrylic Exterior Primer

Formula No. 1

(Unmodified)

	lb	gal
High speed mill dispersion in order of addition		
Water	167	20.00
Triton N-101	3	0.30
Lecithin WD	2	0.25
Phenylmercuric Acetate (30% sol'n.)	2	0.10
Carbowax 4000	30	3.00
Mix until **Carbowax** is in solution.		
Cellosize QP-4400	2	0.17
Mix until **Cellosize** hydrates.		
Dee-Fo 495	2	0.30
Ti Pure R-900	150	4.37
Wet Ground Mica (325 mesh)	25	1.04
Barytes #1	50	1.36
Disperse for 10–15 min, then reduce speed and add:		
Rhoplex AC-235	440	50.00
Tributyl Phosphate	8	1.00
Triton X-405	5	0.50
Dee-Fo 495	2	0.30
Tanblok TP-4 } Premix	10	0.80
Water } Premix	142	17.00

Add **Tanblok** to water stirring continuously until solution is clear, about 10 min. Add solution to batch.

Note 1

The **Tanblok TP-4** solution should be added to the batch as rapidly as possible. If added too slowly, the paint may body momentarily. This structuring can be broken by increasing the speed of the Cowles dissolver or by adding the balance of the solution all at once.

Note 2

After the paint is completed, mixing should be continued for several minutes to insure the incorporation of any structured paint.

Note 3

Initially the paint has Newtonian flow, however, upon standing it will re-structure. Simple stirring or shaking will break the structure immediately.

PVC (%)	22.7
NV (%)	45.6
pH	7 ± 0.5

No. 2

(Emulsion)

	lb	gal
Ti Pure R-900	100	3.0
Basic Silicate White Lead	50	1.5
Drikalite	75	3.3
Daxad 30	5	0.5
Colloid 606	2	0.3
Ethylene Glycol	28	3.0
Troysan PMA-30	4	0.4
Igepal CO-610	2	0.2
Natrosol 250 H.R. (3%)	50	6.0
Water	73	8.8
Rhoplex AC-34 (46 NV)	650	73.0

PVC (%)	20.0
Visc. (KU)	70–75
Total NV by Vol. (%)	39.1
Total NV by Wt. (%)	50.8
Vehicle NV by Wt. (%)	36.9

This primer can be used on new or unpainted wood surfaces.

Latex Exterior Primer

Formula No. 1

(Stain Resistant)

Water	200.0
Daxad 30	5.0
Triton X102	1.0
Nopco NXZ	1.0
Ethylene Glycol	25.0
Pine Oil	3.0
PMA-60	2.0
Natrosol 250MR	3.5
Titanox RA-NC	200.0
Snowflake	115.0
E-P 202 Lead Silicate	125.0
ParCryl 300	512.0
Nopco NXZ	1.0

PVC (%)	34.7
Total Solids (%)	56.6
Visc. (KU)	80-86
pH	8.7-9.3

No. 2

(Alkyd Modified)

lb/100 gal

High speed mill dispersion in order of addition:

Water	165.0
Troysan CMP-Acetate	0.5
Ethylene Glycol	32.0
Natrosol-250 H.R.	3.0

Mix at high speed until cellulose is completely hydrated. Reduce speed and add:

Dee-Fo 233	1.0
Igepal CO-630	2.0

Tamol-850	7.0
Strodex MOK-70	3.0
Dowicil S-13	6.0

Increase to high speed and add:

Ti Pure R-960	150.0
Kaopaque 30	130.0
Zinc Oxide (N.J. xx-601)	50.0
Troysan PMO-30	1.0
Water	65.0

Disperse for 10–15 min, then add:

Ucar-366		365.0
Cargill 7700	Premix before adding	65.0
Advacar (6% cobalt drier)	Premix before adding	1.0
Advacar (6% zirco drier)	Premix before adding	1.0
Dee-Fo 233		1.5
Igepal CO-630		3.0
Water and/or **Natrosol-250 H.R.** (2% sol'n.)		60.0

PVC (%)	28.5
lb/gal	11.1
Visc. (KU)	80–85
Total NV by Wt. (%)	55.5

No. 3

(Stain-Resistant)

Charge into mixer under agitation:

	lb	gal
Water	165.0	19.8
Cosan 265	1.0	0.1
Cellosize QP-4400	3.5	0.4
Ethylene Glycol	15.0	1.6
Igepal CTA-639	2.0	0.2
Foamaster R	1.0	0.1
Strodex MOK-70	3.0	0.3

Tamol 850	7.0	0.8
Dowicil S-13	6.0	0.5
Ti Pure R-960	150.0	4.5
Halox CX-15	130.0	5.2
Zinc Oxide (N.J. xx-503)	50.0	1.1

Disperse in high-speed mill, then add slowly with agitation:

Water	80.5	9.7
Ucar Latex 2345 or **366**	365.0	40.0
Cargill 7700	65.0	7.8
Advacar (6% cobalt drier)	1.0	0.1
Advacar (5% zirco drier catalyst)	1.0	0.1
Foamaster R	1.5	0.2
Igepal CTA-639	3.0	0.3
Water	60.0	7.2

Note:
Equivalent grades of TiO_2, zinc oxide, and defoamer may be substituted, if desired, with minimum change in properties.

PVC (%)	28.5
lb/gal	11.1
Visc. (KU)	75–80
Total Solids (%)	55.5

Styrene-Acrylic Wood Primer

	Formula No. 1	No. 2
Tioxide R-CR2	250.0	250.0
Microfine P	107.0	107.0
Microcal ET	10.0	10.0
Vinapol 1640 (diluted 1:6 with water)	35.0	35.0
Potassium Hydroxide (50% aq. sol'n.)	2.0	2.0
Celacose CP10 (2% aq. sol'n.)	90.0	90.0
Water	127.0	117.6
Foamaster NXZ	2.0	2.0
Acticide MPM	0.4	0.4

Dalpad A	10.0	20.0
Vinamul 3240	366.0	–
Vinacryl 7172	–	366.0
PVC (%)	40	40
Total NV (%)	55	55

Acrylic Latex Metal Primer

	lb/100 gal
Zinc Yellow	333
Titanium Dioxide	41
Magnesium Silicate	40
Phthalocyanine Blue	1
Bentone 38	3
Soya Lecithin	2
Pliolite S-5A	146
Chlorinated Paraffin (40% chlorine)	16
Xylene	240
VM&P Naphtha	208
Methanol (methyl alcohol)	1

Alkyd Primer

Formula No. 1

(Rust-Inhibitive, Air-Dry)

	lb	gal
Arolon 376-WA8-50	181.9	20.91
Triethylamine	2.2	0.36
Red Iron Oxide	72.8	1.70
Strontium Chromate	36.4	1.16
Unbleached Micronized Barytes	218.3	5.96
Micronized Talc	36.4	1.62
Deionized Water	206.9	24.84
Drew L-475	1.5	0.19

Pebble mill 48 h.

Arolon 376-WA8-50	181.9	20.91
Lead Emulsive Drier (24%)	0.4	0.04
Cobalt Emulsive Drier (6%)	1.6	0.20
Deionized Water	256.0	30.73
Exkin No. 2	1.0	0.13

Visc. (s—#4 Ford Cup)	160–180
pH	7.5–8.0

No. 2

(Black Dipping)

	lb	gal
Bentone 27	7.2	0.51
Isopropanol	7.2	1.10
Deionized Water	34.0	4.09

Disperse in mill 15 min.

Arolon 585-W-43	185.0	20.77
Nonbeaded Carbon Black	14.4	0.96
Oncor M5	46.1	1.38
Unbleached Barytes	136.5	3.75
Lo Micron 399	93.6	4.18
Triton X-405	6.9	0.75
Cobalt Naphthenate (6%)	2.0	0.26
Lead Naphthenate (24%)	0.5	0.05
Butoxy Ethanol	9.5	1.26
Potassium Tripolyphosphate	1.3	0.13
Drew L-475	1.7	0.23
Deionized Water	114.8	13.78

Pebble mill to 6¼ grind

Arolon 585-W-43	392.0	44.04
Deionized Water	23.0	2.76

P/B Ratio	1.2
Total NV (%)	49.5
lb/gal	10.75
Visc. (s—#2 Zahn Cup)	≈ 18–25

No. 3

(Black)

	lb	gal
Arolon 363-WA8-50	142.4	16.47
KTPP	1.1	0.05
United 3017L	10.6	0.71
Sparmite	159.5	4.36
Lo Micron 399	42.5	1.89
Busan 11-M1	53.2	1.91
Triethylamine	0.5	0.09
Water	159.5	19.16

Pebble mill 18 h:

	lb	gal
Arolon 363-WA8-50	212.6	24.58
Manganese Naphthenate (6%)	0.6	0.09
n-Butanol	5.3	0.79
Water	248.7	29.90

Total NV (%)	42.9
P/B by Wt. (%)	1.5/1.0
Visc. (s—#4 Ford Cup)	40–60
pH	8.0–8.8

No. 4

(Air)

	lb	gal
Arolon 969 (30%)	299.6	34.76
R-2199	180.0	4.19
Oncor M50	90.0	2.64
Sparmite	45.0	1.22
Talc—Platey Western Type	45.0	2.01
Ben-A-Gel EW	3.6	0.25
Water	103.5	12.43

Pebble mill 18 h.

	lb	gal
Arolon 585-W43	107.1	12.03
Water	72.0	8.66

Premix the following and add slowly with agitation to **Arolon 585**.

Arolon 969 (30%–ammonia neutralization)	150.3	17.43
Mn Naphthenate (6%)	1.5	0.19
Co Naphthenate (6%)	1.5	0.19
n-Butanol	27.0	4.00

Total NV (%)	48.4
P/B Ratio by Wt.	2.0/1.0
Visc. (s–#4 Ford Cup)	50–70
pH	7.5–8.5

No. 5

(Red)

	lb	gal
Arolon 585-W-43	125.26	14.40
R-2199	52.40	1.22
Lampblack	1.05	0.07
Oncor M50	78.59	2.31
Primex	104.79	2.88
Lo Micron 399	52.40	2.33
Deionized Water	34.93	4.19

Pebble mill 18 h and add

Arolon 525-W-37	62.60	7.20

Mix 15 min and let down with:

Arolon 525-W-37	438.03	50.35
Propanol B	56.59	7.73
Manganese Naphthenate (6%)	2.10	0.27
Deionized Water	58.68	7.05

Total NV (%)	48.8
P/B by Wt.	1.25
Visc. (s–#4 Ford Cup)	30–50

No. 6

(Weldable)

Phenolic Modified Alkyd (50% in xylene)	19.40
M-P-A (mineral spirits)	1.95
Xylene	18.84
Aluminum Paste (74% in mineral spirits)	2.65
Cobalt Naphthenate	0.38
Manganese Naphthenate	0.40
Lithium Oxide	0.10
Silica Gel	0.47
Zinc Dust (avg. size 6 μ)	55.81

No. 7

(Zinc Phosphate, Anticorrosive)

	lb
Zinc Phosphate	270.0
1681 Red Oxide	97.0
Nytal 400	82.0
Medium Soya Alkyd (1331-50 soya)	303.0
Mineral Spirits	77.0

Disperse in Hockmeyer. Add:

Medium Soya Alkyd	194.0
Mineral Spirits	26.0
Xylol	6.0
Troykyd Cobalt Naphthenate (6%)	1.5
Troykyd Lead Naphthenate (24%)	5.8
Troykyd Antiskin	1.4

Chlorinated Rubber Metal Primer

Formula No. 1

(Aluminum, Low-Build)

Chlorinated Rubber (20 cps)	18.7
Chlorinated Paraffin 42	8.0

Aluminum Paste (65% nonleafing)	24.2
Zinc Oxide	0.9
Mixed Trimethyl Benzenes	48.2

No. 2

(Aluminum/Basic Lead Sulfate–High-Build)

Chlorinated Rubber (10 cps)	12.0
Chlorinated Paraffin 42	3.9
Chlorinated Paraffin 70	6.8
Red Iron Oxide	7.7
Basic Lead Sulfate	15.2
Barytes	7.7
Aluminum Paste (65% standard nonleafing)	10.5
Modified Hydrogenated Castor Oil	0.5
Solvent Blend of Xylene and Mixed Trimethyl Benzenes at 4:1 Ratio	34.7

No. 3

(For Hopper and Tank Cars)

Primer Type	Parlon-All Chlorinated
Parlon S-10	13.8
Aroclor 1254	5.4
Aroclor 5460	3.4
Zinc Yellow	6.0
Micro-Mica C-3000	6.3
Zinc Oxide	3.0
Asbestine 325	6.3
Rutile Titanium Dioxide	2.3
Thixatrol ST	1.1
Toluene	39.0
Xylene	13.0
Epichlorohydrin	0.1
ERL-2774	0.3

Solids, as prepared (%)	47.9

Application:
To maintain maximum chemical resistance under an all-chlorinated top coat, an all-chlorinated primer is suggested. The primer should be applied to a dry film thickness of two mils in one coat. Two coats of the top coat are applied to give a total thickness of 5 mils, including primer.

Phenoxy Ship Metal Primer

(Iron Yellow)

This coating is recommended as a fast drying shop coat for sand or shot blasted steel. Recommended dry film thickness is 0.8-1.8 mils. Recommended diluent: 1/1/1: MEK/Toluene/Acetone by weight.

	lb	gal
PKHC, Phenoxy Resin	96.6	10.05
Polyketone 251	24.7	2.55
Light Lemon Yellow Oxide 1101	187.5	5.50
Alcan 5100	14.5	0.65
M-P-A 60 (toluene)	3.9	0.45
MEK	183.5	27.30
Toluene	185.5	25.50
Acetone	185.5	28.00

Procedure:
Predissolve the PKHC and **Polyketone 251** in the solvent blend. Thoroughly disperse the pigments in the resin solution on a high speed mixer (e.g., Cowles dissolver).

Visc. (cps)	60–70
Total NV (%)	36.93
P/B Ratio (%)	32.99
lb/gal	8.81

PVC Metal Primer

Formula No. 1

	lb	gal
Charge the following under agitation:		
Water	131.00	15.80
Calgon (unadjusted–10% sol'n.)	5.60	0.53
Tamol 731 (25%)	13.40	1.47
Defoamer	6.10	0.87
Mix well, then add under agitation:		
Titanium Dioxide	200.00	5.72
Asbestine 3X	167.50	7.05
Busan 11-M1	134.00	4.84

Grind the above at high speed 20–25 min; lower speed, add the following under agitation:

	lb	gal
Defoamer	3.00	0.44
Polidene P-761	554.48	56.88
Water*	55.54	6.66
Butyl **Carbitol** Acetate	12.20	1.52

lb/gal	12.3–12.7
Visc. (KU)	80–85
PVC (%)	38.0
Total NV by Wt. (%)	63.6
pH	8.4

*Note:

To make an oil modified primer, replace this quantity water with a blend of:

	lb	gal
Castung 235	51.43	6.60
Cobalt (6%)	0.87	0.10

No. 2

(Golden)

	lb	gal
Charge the following under agitation:		
Water	131.00	15.80
Calgon (unadjusted–10% sol'n.)	5.60	0.53
Tamol 731 (25%)	13.40	1.47
Defoamer	6.10	0.87
Mix well, then add under agitation:		
Titanium Dioxide	100.00	2.86
Titanium Golden	100.00	2.93
Asbestine 3X	167.50	7.05
Busan 11-M1	134.00	4.84
Grind the above at high speed, 20–25 min; lower speed, add the following under agitation:		
Defoamer	3.00	0.44
Polidene P-761	554.48	56.88
Water*	55.54	6.66
Butyl **Carbitol** Acetate	12.20	1.52

lb/gal	12.4–12.8
Visc. (KU)	70–75
PVC (%)	38.0
Total NV by Wt. (%)	64.0
pH	8.2

*Note:

To make an oil modified primer replace this quantity water with a blend of:

	lb	gal
Castung 235	51.43	6.60
Cobalt (6%)	0.87	0.10

No. 3

	lb	gal
Charge the following under agitation:		
Water	131.00	15.80
Calgon (unadjusted–10% sol'n.)	5.60	0.53
Tamol 731 (25%)	13.40	1.47
Defoamer	6.10	0.87
Preservative	1.80	0.11
Mix well, then add under agitation:		
Titanium Dioxide	200.00	5.72
Asbestine 3X	167.00	7.05
Baryta X5R	134.00	3.62

Grind the above at high speed (20-25 min), lower speed, add following under agitation:

Defoamer	3.00	0.44
Polidene P-761	554.48	56.88
Water*	55.54	6.66
Butyl **Carbitol** Acetate	12.20	1.52

lb/gal	12.55-12.95
Visc. (KU)	75-80
PVC (%)	38.0
Total NV by Wt. (%)	63.4
pH	7.2

*Note:

To make an oil modified primer, replace this quantity water with a blend of:

	lb	gal
Castung 235	51.43	6.60
Cobalt (6%)	0.87	0.10

Styrene Acrylic Latex Metal Primer

(White, Alkyd)

Charge into mixer under agitation:

		lb	gal
Natrosol 250 HR (1.5% sol'n.)		89.0	10.7
KTPP		1.0	0.1
Tamol 731		4.0	0.4
Triton CF 10		5.0	0.6
Nopco NXZ		1.0	0.1
Ethylene Glycol		20.0	2.1
Ti Pure R 902		125.0	3.7
Duramite		50.0	2.2
Molywhite 212		75.0	3.0
Gold Bond R		50.0	2.2
Davenite P 12		25.0	1.1
Cosan 635 W		1.0	0.1

Disperse in high speed mill, then add slowly with agitation:

		lb	gal
Arolon 585	Premix	100.0	11.1
Super Cobalt Drier	Premix	1.0	0.1
Zirco Drier Catalyst (6%)	Premix	1.5	0.1
Nacrylic 8403 (50%)		468.0	53.8
Ammonium Hydroxide (28%)		1.0	0.1
Nopco NXZ		1.0	0.1
Dowanol DPM		28.0	3.1
Water		43.5	5.2
Strodex PK-90		10.0	1.1

PVC (%)	28.1
lb/gal	11.1
Visc. (KU)	70–75
Total Solids (%)	55.5

Vinyl Primer for Coil Coating

(Steel and Galvanized Steel Substitutes)

	lb	gal
Bakelite VMCC	131.57	11.79
Bakelite BKR 2620	19.74	1.91
Mapleo 387	104.78	2.50
Rutile Titanium Dioxide	17.36	0.51
TCP Plasticizer	19.74	2.10
Lexinol AC-1	1.32	0.17
Bakelite ERL-2774	9.18	0.94
Thermolite 31	3.91	0.49
M-P-A 60 (xylene)	4.59	0.64
MIBK	282.87	42.41
Toluene	263.16	36.54

Procedure:

Dissolve resins **VMCC** and **BKR-2620** in solvents MIBK and toluene. Charge to pebble mill: iron oxide, TiO_2, TCP plasticizer, **Lexinol AC-1**, **ERL-2774**, **Thermolite 31**, **M-P-A 60** xylene, and one-half of above resin solution. Grind to a fineness of grind of 7 (Hegman). Let down with remaining solution.

Visc. (s—#4 Ford Cup)	100-110
NV (%)	36
P/B Ratio	0.649

Vinyl Primer

(Basic Lead Silico-Chromate)

This coating is intended for use over (poly) vinyl butyral (**Bakelite** Resin XYHL) metal pretreatment coatings (wash primers) such as VP-2980.

	lb	gal
Bakelite VAGH	123.3	10.67
Basic Lead Silico-Chromate	101.5	2.98
Dibasic Lead Phosphite	6.2	0.12
Pure Red Iron Oxide	7.3	0.17
M-P-A 60 (xylene)	6.3	0.85
Epichlorohydrin	0.9	0.09

MIBK	145.0	21.71
MIAK	145.0	21.39
Toluene	253.8	34.95
Isopropyl Alcohol	29.0	4.43
Flexol Plasticizer TCP	29.0	2.99

lb/gal	8.44
PVC (%)	18.4
Total NV by Wt. (%)	32.3
Total NV by Vol. (%)	17.7

Procedure:

Dissolve vinyl resin in solvents and plasticizer. Combine other ingredients and solution; then disperse to a fineness of 6 on the Hegman gauge.

Polyvinyl Butyral Wash Primer

Unlike most additives checked in this wash primer, MPA-Butanol does not destroy the adhesion of the coating.

	lb	gal
Resin Component:		
Bakelite XYHL	56.0	6.10
Zinc Chromate (insoluble type)	54.0	1.78
Magnesium Silicate (Mil-M-15173, Type A or B)	8.0	0.34
Lampblack	0.6	0.04
n-Butyl Alcohol	125.0	18.48
Isopropyl Alcohol (99%)	353.0	53.80
Water	15.0	1.80
M-P-A (butanol)	0 or 16.0	0 or 2.20
Acid Component:		
Phosphoric Acid	28.0	2.0
Water	25.0	3.0
Isopropyl Alcohol	99.0	15.0

Procedure:

Ball mill at approximately 110 F.

lb/gal	7.45
PVC (%)	21
Visc. (KU)	67

Latex Maintenance Primers

	Formula No. 1 (*Zinc Dust, For Galvanized Steel*)		No. 2 (*Utility, For Shop Coat*)		No. 3 (*Zinc Chromate, For Maintenance*)	
	lb	gal	lb	gal	lb	gal
Add ingredients in order shown:						
Aromatic High Flash Naphtha	115.0	16.000	127.0	17.700	130.0	18.100
Marbon 9200 MV	140.0	16.000	127.0	14.500	130.0	14.800
Chlorinated Diphenyl (solid)	20.0	1.400	–	–	–	–
Mix to clear solution.						
Chlorinated Paraffin (50%)	–	–	37.0	3.6000	38.0	3.700
Raw Linseed Oil	35.0	4.500	37.0	4.800	38.0	4.900
Chlorinated Diphenyl (liquid)	20.0	1.600	–	–	–	–
Talc	–	–	185.0	7.800	80.0	3.400

Zinc Oxide	145.0	3.100	25.0	0.500	50.0	1.700
Zinc Chromate	–	–	–	–	50.0	1.700
Red Iron Oxide	–	–	125.0	3.100	200.0	5.000
Cobalt Tallate (6%)	0.2	0.025	0.2	0.025	0.2	0.025
*Suspension Additive	3.0	0.200	3.0	0.200	3.0	0.200
*Denatured Alcohol	1.5	0.200	1.5	0.200	1.5	0.200

*Prewet.

Mix to a smooth paste and grind on a 3-roll mill to a fineness of 5–6 N. S.

Mineral Spirits (KB40+)	220.0	33.300	223.0	33.800	221.0	33.500
Aromatic High Flash Naphtha	105.0	14.600	96.0	13.300	91.0	12.600
Zinc Dust, to be packaged separately and added just prior to use	580.0	9.900	–	–	–	–

Total Solids (%)	67.9	54.6	57.0
PVC (%)	36.0	33.6	32.8
lb/gal	13.7	9.9	10.4
Visc. (KU)	79	71	71

No. 4

(Alkyd Modified)

		Pt/Wt	Pt/Vol
Water	Premix	165.00	19.800
Natrosol 250 MR (100%)	Premix	1.73	0.155
Nopco NXZ	Premix	0.90	0.115
Ammonium Hydroxide (28%)	Premix	0.87	0.115
Tamol 850 (30%)		13.48	1.350
Triton CF-10		2.12	0.240
Ti Pure R-960		144.40	4.240
Water Ground Mica (325 mesh)		24.07	1.030
Atomite		105.89	4.690
Busan 11-M1		96.26	3.510

Let down:

		Pt/Wt	Pt/Vol
Rhoplex MV-2 (46.0%)	Premix	446.86	50.050
Triton X-405 (70%)	Premix	1.64	0.170
Texanol		3.66	0.460
Aroplaz 1271*	Premix	57.76	6.910
Triton X-100	Premix	4.28	0.480
Ethylene Glycol		22.33	2.410
Nopco NXZ		0.90	0.115
Sodium Nitrite (13.8% in water)		5.78	0.690
Water/250 MR (2.5%)		28.88	3.470

PVC (%)	32.6
Total Solids by Vol. (%)	41.9
Visc.—Initial (KU)	65

Adjust pH to 9.3-9.5 with 28% ammonium hydroxide.

---

* **Aroplaz 1271** drier treated with 0.3 lb of 6% cobalt, 0.3 lb of 6% zirconium, and 3.4 lb of 6% manganese.

No. 5

		Pts./Wt.	Pts./Vol.
Water	Premix	110.10	13.13
Natrosol 250 MR (100%)	Premix	2.00	0.12
Nopco NXZ	Premix	0.90	0.13
Ammonium Hydroxide (28%)	Premix	0.90	0.13
Tamol 731 (25%)		9.30	1.01
Triton X-405		2.30	0.25
Ethylene Glycol		23.40	2.52
Ti Pure R-960		151.50	4.44
Water Ground Mica (325 mesh)		25.25	1.08
Atomite		151.50	6.72
Molywhite 212		50.50	2.01
Kadox		3.00	0.06

Grind the above in a high-speed impeller mill at 3800–4500 ft/min for 10–15 min. At a slower speed, letdown as follows:

	Pts./Wt.	Pts./Vol.
Rhoplex MV-2 (46%)	595.20	66.66
Texanol	4.85	0.63
Skane M-8	2.20	0.25
Troykyd 999	0.90	0.13
Sodium Nitrite (13.8% in water)	6.10	0.73

PVC (%)	33.5
Total Solids by Vol. (%)	42.8
pH	≈ 9.5
Visc. (KU)	≈ 65-68

No. 6

		Pts./Wt.	Pts./Vol.
Water	Premix	159.32	19.12
Natrosol 250 MR (100%)	Premix	1.67	0.15
Nopco NXZ	Premix	0.86	0.11
Ammonium Hydroxide (28%)	Premix	0.84	0.11
Tamol 850 (30%)		13.01	1.30
Triton CF-10		2.04	0.23

Ti Pure R-960	139.43	4.09
Water Ground Mica (325 mesh)	23.24	1.00
Atomite	102.25	4.53
Busan 11-M1	92.95	3.39
Let down:		
Rhoplex MV-2 (46.0%)	556.14	62.29
Texanol	4.55	0.58
Ethylene Glycol	21.57	2.32
Nopco NXZ	0.86	0.11
Sodium Nitrite (13.8% in water)	5.58	0.67

PVC (%)	33.2
Total Solids by Vol. (%)	39.7
Visc.–Initial (KU)	≈ 65

Adjust pH to 9.3–9.5 with 28% ammonium hydroxide.

No. 7

(Alkyd Modified)

		Pts./Wt.	Pts./Vol.
Water	Premix	161.48	19.38
Natrosol 250 MR (100%)	Premix	1.70	0.15
Nopco NXZ	Premix	0.88	0.11
Ammonium Hydroxide (28%)	Premix	0.85	0.11
Tamol 850		13.19	1.32
Triton CF-10		2.07	0.24
Ti Pure R-960		141.32	4.15
Water Ground Mica (325 mesh)		23.55	1.01
Atomite		103.64	4.59
Busan 11-M1		94.22	3.44
Let down:			
Rhoplex MV-9 (45.5%)	Premix	437.35	50.29
Triton X-405	Premix	1.60	0.17
Texanol		10.08	1.27

Aroplaz 1271* } Premix	56.53	6.76
Triton X-100 }	4.19	0.47
Ethylene Glycol	21.86	2.36
Nopco NXZ	0.88	0.11
Sodium Nitrite (13.8% in water)	5.65	0.68
Water/250 MR (2.5%)	28.26	3.39
PVC (%)	31.7	
Total Solids by Vol. (%)	41.9	
Visc.–Initial (KU)	≈ 65	

Adjust pH to 9.3–9.5 with 28% ammonium hydroxide.

---

* **Aroplaz 1271** drier treated with 0.3 lb of 6% cobalt, 0.3 lb of 6% zirconium, and 3.4 lb of 6% manganese.

No. 8

	Pts./Wt.	Pts./Vol.
Water	155.23	18.63
Natrosol 250 MR (100%)	1.63	0.14
Nopco NXZ	0.84	0.11
Ammonium Hydroxide (28%)	0.82	0.11
Tamol 850 (30%)	12.68	1.27
Triton CF-10	1.99	0.23
Ti Pure R-960	135.84	3.98
Water Ground Mica (325 mesh)	22.64	0.97
Atomite	99.62	4.41
Busan 11-M1	90.56	3.31

Let down:

Rhoplex MV-9 (45.5%)	541.84	62.31
Texanol	11.95	1.51
Ethylene Glycol	21.01	2.26
Nopco NXZ	0.84	0.11
Sodium Nitrite (13.8% in water)	5.43	0.65

PVC (%)	32
Total Solids by Vol. (%)	39.5
Visc.–Initial (KU)	65

Adjust pH to 9.3–9.5 with 28% ammonium hydroxide.

No. 9

	Pts./Wt.	Pts./Vol.
Water	97.06	11.65
Natrosol 250 MR (100%)	0.98	0.06
Nopco NXZ	0.91	0.12
Ammonium Hydroxide (28%)	0.88	0.12
Emultex R	2.05	0.25
Surfactant X-QS-20	1.97	0.21
Triton CF-10	21.5	0.25
AMP-95	2.94	0.38
Ethylene Glycol	32.56	3.51
Water Ground Mica (325 mesh)	24.52	1.04
Atomite	220.62	10.00
Kroma Red RO-8097	49.02	1.18
Molywhite 212	49.02	1.96
Kadox 515	11.76	0.26

Grind the above in a high speed impeller mill at 3800–4500 fpm for 10–15 min. At a slow speed, letdown as follows:

Rhoplex MV-9 (45.5%)	455.15	52.33
Linseed Oil (pure raw)	55.59	7.22
Nopco NXZ	0.91	0.12
Skane M-8	2.12	0.25
Texanol	10.47	1.33
Sodium Nitrite (13.8% in water)	5.96	0.71
Water/**Natrosol 250 MR**	53.39	6.40
Cobalt Drier (6%)	0.25	0.03
Zirconium Drier (6%)	4.42	0.62

PVC (%)	32.7
Total Solids by Vol. (%)	44.55
pH–Initial	9.5
Visc.–Initial (KU)	68

Styrene Acrylic Latex Maintenance Primer

(Brown)

Cellosize QP-4400 (3% sol'n.)	102
KTPP	1
Strodex PK-90	3
Triton CF-10	3
Nopco NDW	1
B-2093F (brown iron oxide)	166
Oncor F-31	47
W.G. Mica (325 mesh)	24
Vicron 15-15	142
Barytes W-1430F (unbleached)	47
Ethylene Glycol	19
Texanol	25
Water	75

Cowles for 15–20 min

Resyn 8403	510
Nopco NDW	1
Ammonium Hydroxide (28%)	1

PVC (%)	39
Visc. (KU)	75–80

Chapter II

EXTERIOR PAINTS

Paints with a low pigment load usually have exterior durability. The large binder content provides the film matrix required to withstand exterior conditions. The pigment volume concentration (PVC) is usually well under 40 percent.

The low PVC (i.e. high binder) results in a strongly bound, highly resistant film possessing the following variable qualities: good weathering, glossiness, resistance to specific chemicals and abrasion. The formulas presented in this chapter possess one or more of these characteristics to a greater or lesser degree.

Exterior Acrylic Emulsion Paint

Formula No. 1

		lb	gal
Water		200.0	24.00
Nopco NXZ		1.0	0.13
Ethylene Glycol	Premix	27.9	3.00
Hydroxyethyl Cellulose	Premix	2.0	–
Tamol 850 (30%)		10.5	1.05
KTPP		1.5	–
Triton CF-10		2.5	0.28
Rutile Titanium Dioxide		250.0	7.60
Zinc Oxide		25.0	0.54
Talc		175.0	7.35

Grind the above materials in a high speed mill (Cowles, 3800–4200 fpm for 10–15 min) then add the following at slower speed:

Rhoplex AC-388 (50%)	440.0	50.00
Nopco NXZ	1.0	0.13
Texanol	11.0	1.39
Skane M-8	0.5	0.06
Water and/or HEC (2.5% sol'n.)	37.2	4.47

PVC (%)	39.7
Total Solids by Vol. (%)	39.0
Total Solids by Wt. (%)	56.5
Visc. (KU)	77
pH	9.3–9.5

	No. 2 *(White)*		No. 3 *(Tint Base)*	
	lb	gal	lb	gal
Ti Pure R-900	200	5.9	150	4.4
Drikalite	350	15.6	200	8.9
Daxad 30	5	0.5	5	0.5
Colloid 606	2	0.3	2	0.3
Ethylene Glycol	28	3.0	28	3.0
Troysan PMA-30	4	0.4	4	0.4
Igepal CO-610	2	0.2	2	0.2
Natrosol 250 HR (3%)	75	9.0	92	11.0
Water	126	15.1	119	14.3
Rhoplex AC-35 (46 NV)	440	50.0	501	57.0

	No. 2	No. 3
PVC (%)	50.0	35.0
Visc. (KU)	80–85	80–85
Total Paint NV by Vol. (%)	43.0	38.0
Total Paint NV by Wt. (%)	60.5	52.7
Total Vehicle NV by Wt. (%)	30.2	31.2

	No. 4 (White) lb	No. 4 (White) gal	No. 5 (Tint Base) lb	No. 5 (Tint Base) gal
Ti Pure R-900	200	5.9	150	4.4
Drikalite	350	15.6	200	8.9
Daxad 30	5	0.5	5	0.5
Colloid 606	2	0.3	2	0.3
Ethylene Glycol	28	3.0	28	3.0
Troysan PMA-30	4	0.4	4	0.4
Igepal CO-610	2	0.2	2	0.2
Natrosol 250 HR (3%)	58	7.0	83	10.0
Water	175	21.1	161	19.3
Rhoplex AC-388 (50 NV)	405	46.0	466	53.0
PVC (%)	50.0		35.0	
Visc. (KU)	80–85		80–85	
Total Paint NV by Vol. (%)	43.0		38.0	
Total Paint NV by Wt. (%)	61.4		53.0	
Total Vehicle NV by Wt. (%)	30.3		31.5	

No. 6

Water		200.0
Defoamer*		1.0
Ethylene Glycol	Premix	27.9
Hydroxyethyl Cellulose	Premix	2.0
Tamol 850 (30%)		10.5
KTPP		1.5
Triton CF-10		2.5
Rutile Titanium Dioxide		250.0
Zinc Oxide		50.0
Talc		150.0

Grind the above materials in a high-speed mill (Cowles, 3800–4200 fpm for 10–15 min) then add the following at a slower speed as follows:

Rhoplex AC-64 (60.5%)	363.6
Defoamer*	1.0
Texanol	11.0

Skane M-8	1.5
Water and/or Hydroxyethyl Cellulose (2.5% sol'n.)	115.9

PVC (%)	39.1
Total Solids by Vol. (%)	38.7
Total Solids by Wt. (%)	56.5
Visc. (KU)	80
pH	9.3-9.5

* **Nopco NXZ** has been found satisfactory with **Rhoplex AC-388** while **Foamaster VL** is suggested for use with **Rhoplex AC-64.**

Note:

Toning to a reflectance of 87% is required to attain the specified dry opacity in both formulations. 0.1 lb **Cal Ink** 8800 line E Phthalo Blue plus 0.1 lb **Cal Ink** 8800 line B Lamp Black is used in these preparations.

Acrylic House Paint

(Alkyd Modified)

	≈lb/100 gal
Horse Head R760	292
Calcium Carbonate	68
Gold Bond Silica	74
Zinc Oxide XX503	50
HEC (4½%)	59
Water	147
KTPP	5
Strodex MOK-70	5
Igepal CO-630	3
Colloids 518-B	2
Ethylene Glycol	25
Propylene Glycol	35
Tributyl Phosphate	12
Ammonia (28%)	2

Attagel 40	10
Rhoplex AC-388	370
Aroplaz 1271 (100%)*	46

* Contains driers

lb/gal	12.1
PVC (%)	40.0

Acrylic Latex Exterior Paint

Formula No. 1

Natrosol 250 MR	144.3
Ethylene Glycol	25.0
Water	15.0
Victawet LPF	3.6
KTPP	1.5
Triton CF-10	2.5
Drew L-475	1.0
Propylene Glycol	34.0
Ti Pure R-960	250.0
AZO-11 ZNO	50.0
Asbestine 3X	178.9
Rhoplex AC-64	380.0
Drew L-475	3.0
Texanol	11.0
Skane M-8	2.0
Ammonium Hydroxide	2.0
Water	106.9
lb/gal	12.06
Visc.—Initial (KU)	77

No. 2

Natrosol 250 MR	144.3
Ethylene Glycol	25.0
Water	17.5

Victawet LPF	2.7
KTPP	1.5
Triton CF-10	2.5
Drew L-475	1.0
Propylene Glycol	34.0
Ti Pure R-960	250.0
AZO-11 ZNO	50.0
Asbestine 3X	178.9
Rhoplex AC-64	380.0
Drew L-475	3.0
Texanol	11.0
Skane M-8	2.0
Ammonium Hydroxide	2.0
Water	105.3

lb/gal	12.10
Visc.–Initial (KU)	78

Ethyl Vinyl Acetate Exterior Paint

Formula No. 1

Natrosol 250 MR	225.0
Water	100.0
Victawet LPF	10.1
Igepal CO-G10	3.0
Ethylene Glycol	30.0
Texanol	2.0
Drew L-475	2.0
Horse Head RF-31	250.0
Horse Head XX-503	50.0
Snowflake	75.0
Dowicil S-13A	3.0
Water	65.9
Drew L-475	2.0
Aroplaz 1271	38.0

Nuxtra Zirconium		0.7
Nuxtra Cobalt		1.5
Airflex 728		288.0
lb/gal	11.33	
Visc.–Initial (KU)	90	

No. 2

Natrosol 250 MR		225.0
Water		100.0
Victawet LPF		8.1
Igepal CO-G10		3.0
Ethylene Glycol		30.0
Texanol		5.0
Drew L-475		2.0
Horse Head RF-31		250.0
Horse Head XX-503		50.0
Snowflake		75.0
Dowicil S-13A		3.0
Water		67.9
Drew L-475		2.0
Aroplaz 1271		38.0
Nuxtra Zirconium		0.7
Nuxtra Cobalt		1.5
Airflex 728		288.0
lb/gal	11.31	
Visc.–Initial (KU)	91	

Latex Exterior Paint

Formula No. 1

(Low Cost, Semigloss)

	lb	gal
Charge the following to the tank:		
Tamol 731 (25%)	7.2	0.78
Propylene Glycol	20.0	2.32
Water	30.0	3.60

Hydroxyethyl Cellulose–2.5% **(QP-4400)**	92.0	11.04
Colloid 600	2.0	0.26

Add the following at low speed:

Rutile Titanium Dioxide (high gloss)	170.0	4.96
Delore X5R (supermicrofine)	75.0	2.02

Grind the above in a high-speed mill (Cowles, 3800–4500 fpm for 10–15 min) and let down at lower speed with the following:

Rhoplex AC-490 (46.5%)		382.0	43.16
Texanol	Premix	10.0	1.26
Triton GR-7	Premix	2.0	0.23
Super Ad-It	Premix	0.5	0.06
Water	Premix	20.0	2.40
Ethylene Glycol	Premix	20.0	2.16
Water and/or Hydroxyethyl Cellulose–2.5% **(QP-4400)***		213.5	25.62
Colloid 600		1.0	0.13

Tone with approximately 0.4 lb of Cal Ink B/100 gal to a reflectance of 82%.

Note:

This formulation is an acceptable tint base formulation for 2 fl oz of Cal Ink 8800 Series colorants if 8 lb of **Triton N-57** are added per 100 gal of paint.

---

* Adjust viscosity to approximately 67 KU.

PVC (%)	27.6
Total Solids by Vol. (%)	25.3
Total Solids by Wt. (%)	40.6
Visc.–Initial (KU)	65–69

No. 2

(Gloss)

Water	Premix until thickening begins	55.1
Hydroxyethyl Cellulose (100%)	Premix until thickening begins	1.2
Ammonium Hydroxide (28%)	Premix until thickening begins	1.5

Tamol 165	16.7
Triton CF-10	2.4
Nopco NXZ	1.2
Ethylene Glycol	38.8
Zopaque RCL-9	256.1
Kadox 515	5.8

Grind the above in a high speed impeller mill at 3800–4500 fpm for 10–15 min. At a slow speed, let down as follows:

Rhoplex MV-9 (45.5%)	619.8
Texanol	28.1
Water	28.1
Troykyd 999	1.2
Skane M-8	1.9
Ammonium Hydroxide (28%)	3.9

PVC (%)	20.0
Total Solids by Vol. (%)	38.2
pH	≈ 9.5
Visc. (KU)	≈ 74

No. 3

(With Triamino Ester, Nonchalking)

	lb/100 gal	gal
Water	150.0	18.00
Ad-A-Blend PD-919	50.0	5.71
Tetrapotassiumpyrophosphate	1.5	0.08
Colloid #677	3.0	0.41
Metasol 285	5.0	0.30
Kelecin FB	4.0	0.46
R-960	225.0	6.75
W.G. Mica (325 mesh)	25.0	1.08
ZZZ-55	50.0	1.07
Camel Carb	75.0	3.32
Cellosize QP-4400, H	4.5	0.39

Disperse

Water	151.8	18.22
PD-109 Latex	375.0	41.21
Chemacoil TA-100	23.4	3.00

lb/gal	11.43
Total NV (%)	54.7
PVC (%)	32.2
Visc. (KU)	78

Procedure:

Formulated for a high speed disperser. Add in order while mixing at a moderate speed. Disperse at 400–4500 fpm. Add last three items in order while stirring at moderate speed. Mix thoroughly.

No. 4

(With Triamino Ester, Self-Cleaning)

	lb/100 gal	gal
Water	175.00	21.00
Ad-A-Blend PD-919B	50.0	5.65
Tetrapotassiumpyrophosphate	1.5	0.08
Colloid 677	3.0	0.41
Kelecin FB	4.0	0.46
Dowicil 100	1.0	0.10
R-960	175.0	5.25
Ti Pure FF	50.0	1.55
ZZZ-55	100.0	2.14
W.G. Mica (325 mesh)	25.0	1.08
Camel Carb	50.0	2.22

Mix

Cellosize QP-4400 H	4.5	0.39
Dowicil S-13	2.5	0.16

Disperse

Water	127.4	15.30
PD-109 Latex	375.0	41.21
Chemacoil TA-100	23.4	3.00
lb/gal	11.67	
Total NV (%)	55.75	
PVC (%)	32.1	
Visc. (KU)	85	

Procedure:

Formulated for a high speed disperser. Add in order while stirring at a moderate speed. Disperse at 4000–4500 fpm. Reduce speed, add last three items, mix thoroughly.

	No. 5 *(White With Alkyd)*	No. 6 *(Tint Base With Alkyd)*
Charge into mixer under agitation:		
Water	65.0	65.0
Merbac 35	0.5	0.5
Ethylene Glycol	35.0	35.0
Strodex MOK-70	5.0	5.0
Tamol 731	7.5	5.5
Igepal CTA-639	3.0	3.0
Colloid 677	1.0	1.0
Cellosize QP-4400 (1½% sol'n.)	165.0	165.0
Metasol TK100	1.0	1.0
Ti Pure R-902	250.0	200.0
Ti Pure PC (Anatase or LW)	25.0	–
St. Joe No. 17 (zinc oxide)	75.0	75.0
Optiwite	35.0	35.0
Minex 4	–	50.0
Celite 281	25.0	25.0

Disperse in high-speed mill, then add slowly with agitation:

Water	30.0	30.0
Ucar Latex 366	271.0	271.0
Igepal CTA-639	6.5	6.5
Advacar (6% cobalt drier)	3.0	3.0
Advacar (6% zirco lead drier)	1.5	1.5
Aroplaz 1271	37.0	37.0
Colloid 677	1.5	1.5
Carbitol Acetate	8.0	8.0
Cellosize QP-4400 (1½% sol'n.)	51.0	63.0
Water	58.5	43.5
PVC (%)	40.0	40.0
lb/gal	11.6	11.3
Visc. (KU)	80-85	80-85
Total Solids by Wt. (%)	53.5	52.7

Polyvinyl Acetate Exterior Paint

Formula No. 1

(Alkyd Modified)

Water	100.0
Natrosol 250 HR	100.0
Ethylene Glycol	28.0
Victawet LPF	7.1
Nopco NXZ	2.0
Titanox 2030	225.0
AZO-33 ZNO	25.0
Asbestine 3X	150.0
Dowicil S-13	3.0
Water	57.9
Natrosol 250 HR	110.0
Igepal CO-630	5.0
Texanol	8.0
Polyco 2358	273.0

Aroplaz 1271		50.0
Nuxtra Zirconium		1.0
Nuxtra Cobalt		1.0
lb/gal	11.38	
Visc.–Initial (KU)	96	

No. 2

(Alkyd Modified)

Water		100.0
Natrosol 250 HR		100.0
Ethylene Glycol		28.0
Victawet LPF		5.7
Igepal CO-630		0.6
Nopco NXZ		2.0
Titanox 2030		225.0
AZO-33 ZNO		25.0
Asbestine 3X		150.0
Dowicil S-13		3.0
Water		59.3
Natrosol 250 HR		110.0
Igepal CO-630		4.4
Texanol		8.0
Polyco 2358		273.0
Aroplaz 1271		50.0
Nuxtra Zirconium		1.0
Nuxtra Cobalt		1.0
lb/gal	11.36	
Visc.–Initial (KU)	95	

No. 3

(Alkyd Modified)

Water	100.0
Natrosol 250 HR	100.0
Ethylene Glycol	28.0

Tamol 850	5.0
Victawet LPF	4.0
Nopco NXZ	2.0
Titanox 2030	225.0
AZO-33 ZNO	25.0
Asbestine 3X	150.0
Dowicil S-13	3.0
Water	72.0
Natrosol 250 HR	94.0
Igepal CO-630	5.0
Texanol	8.0
Polyco 2358	273.0
Aroplaz 1271	50.0
Nuxtra Zirconium	1.0
Nuxtra Cobalt	1.0

lb/gal	11.28
Visc.–Initial (KU)	92

Styrene Acrylic Exterior Coating

(Roller Applied)

Courlose F350 (5% aq. sol'n.)	33.0
Sodium Benzoate	4.0
Tioxide R-CR2	50.0
Fordacal M2	180.0
Fordacal 400	170.0
Fordacal Z3	174.6
Fordacal Z5	120.0
Butyl Dioxitol	10.0
Acticide MPM	0.4
Vinacryl 7172	255.0
Potassium Hydroxide (50% aq. sol'n.)	1.0
Foamaster NDS	2.0

Vinyl Acrylic Exterior-Interior Paint

Formula No. 1

(High Solids)

Kronos RN56	300.0
Whiting BWF42	177.0
Modocoll E1200 (4% aq. sol'n.)	125.6
Calgon S (5% aq. sol'n.)	25.0
Vinapol 1640 (diluted 1:6 with water)	25.0
Sodium Benzoate (10% aq. sol'n.)	10.0
Potassium Hydroxide (50% aq. sol'n.)	2.0
Butyl Dioxitol Acetate	10.0
Nopco Foamaster NXZ	2.0
Acticide MPM	0.4
Vinamul 6705	323.0

PVC (%)	50
Total NV (%)	64
Visc. at 25 C (cps)	≈ 1200
pH	≈ 7

This formula is designed for markets, e.g., in continental Europe, where it is general practice to add water to paints before use; it provides high viscosity and high nonvolatile content.

No. 2

(High Viscosity)

Kronos RN56	270.0
Microdol 1	65.0
Whiting BWF42	33.0
Talc AT1	30.0
Modocoll E1200 (4% aq. sol'n.)	190.0
Calgon S (5% aq. sol'n.)	15.0
Vinapol 1640 (diluted 1:6 with water)	15.0
Sodium Benzoate (10% aq. sol'n.)	10.0
Potassium Hydroxide (50% aq. sol'n.)	2.0
Butyl Dioxitol Acetate	10.0

Foamaster NXZ	2.0
Acticide MPM	0.4
Water	93.6
Vinamul 6705	264.0

PVC (%)	50
Total NV (%)	52
Visc. at 25 C (cps)	≈ 900
pH	≈ 7.5

This system, also, is designed for dilution before use, but provides high viscosity at a more conventional nonvolatile content.

Vinyl-Acrylic Exterior Emulsion Paints

	Formula No. 1 *(White)*		No. 2 *(Tint Base)*	
	lb	gal	lb	gal
Ti Pure R-900	200	5.9	150	4.4
Drikalite	350	15.6	200	8.9
Daxad 30	5	0.5	5	0.5
Colloid 606	2	0.3	2	0.3
Ethylene Glycol	28	3.0	28	3.0
Troysan PMA-30	4	0.4	4	0.4
Igepal CO-610	2	0.2	2	0.2
Natrosol 250 HR (3%)	75	9.0	108	13.0
Water	192	23.1	177	21.3
Resyn 2243 (55 NV)	382	42.0	437	48.0

	No. 1	No. 2
PVC (%)	50.1	35.2
Visc. (KU)	80-85	80-85
Total Paint NV by Vol. (%)	42.9	37.8
Total Paint NV by Wt. (%)	61.4	53.0
Total Vehicle NV by Wt. (%)	30.7	31.8

	No. 3 (White) lb	No. 3 (White) gal	No. 4 (Tint Base) lb	No. 4 (Tint Base) gal
Ti Pure R-900	200	5.9	150	4.4
Drikalite	350	15.6	200	8.9
Daxad 30	5	0.5	5	0.5
Colloid 606	2	0.3	2	0.3
Ethylene Glycol	28	3.0	28	3.0
Troysan PMA-30	4	0.4	4	0.4
Igepal CO-610	2	0.2	2	0.2
Natrosol 250 HR (3%)	75	9.0	107	12.9
Water	192	23.1	170	20.4
Ucar 360 (55 NV)	378	42.0	441	49.0
PVC (%)	50.0		34.6	
Visc. (KU)	80–85		80–85	
Total Paint NV by Vol. (%)	43.1		38.5	
Total Paint NV by Wt. (%)	61.5		53.8	
Total Vehicle NV by Wt. (%)	30.8		32.5	

Acrylic White Exterior Paint

Formula No. 1

(Full Gloss)

Charge the following to a Cowles mixer tank and mix at low speed, ≈ 1500 fpm for 5 min:

	lb	gal
Rhoplex AC-201 (46%)	125.00	13.97
Strodex PK-90 (90%)	3.40	0.36
Triton CF-10	2.00	0.24
SAG-470	0.25	0.03

Slowly add the pigments, increasing the mixer speed to 4200 fpm. Grind for 20 min:

	lb	gal
Ti Pure R-966	248.00	7.44

Gradually add the following and blend at a low speed, ≈ 1500 fpm:

Rhoplex AC-201 (46%)		200.00	22.33
Triethanolamine*	Premix	3.00	0.36
Water	Premix	6.00	0.73
Rhoplex AC-35 (46%)		325.00	36.30

* Additional pH buffer should be added, if required, to stabilize the pH at 9–10.

PVC (%)	20
Total Solids by Wt. (%)	≈ 59
Total Solids by Vol. (%)	≈ 48

No. 2

	lb	gal
Water	162.0	19.4
Igepal CO-630	3.0	0.4
Daxad 30	8.0	0.8
Latekoll D (8%)	40.0	4.7
PMA-30	5.0	0.5
Titanox RANC	225.0	6.4
Titanox A-168-LO	25.0	0.8
Minex 7	100.0	4.6
Nytal 300	50.0	2.1
Deefo 97-2	1.0	0.1

Add in order listed and mix on a high speed disperser for 10 min, then let down as follows:

Ethylene Glycol	37.0	4.0
Texanol	21.0	2.7
Water	100.0	12.0
Acronal 290 D	359.0	41.3
Deefo 97-2	1.0	0.2

PVC (%)	40.0
Total NV by Wt. (%)	54.2
Total NV by Vol. (%)	34.8
lb/gal	11.4
Visc. (KU)	72 ± 5

No. 3

(Emulsion)

	lb	gal
Premix:		
Water	100.0	12.00
Strodex MOK-70	4.0	0.41
Igepal CTA 639	7.0	0.77
KTPP	1.0	–
Increase speed and add slowly:		
Chemcoil TA-100	94.0	12.00
Adjust speed to disperse pigments:		
Ti Pure R-900	240.0	7.13
414W Zinc Oxide	100.0	2.31
Nytal 300	65.0	2.74
Snowflake Whiting	90.0	4.01
Colloid 581-B	1.0	0.13
Decrease speed and add slowly:		
Pine Oil	3.0	0.38
Ethylene Glycol	25.0	2.69
Cellosize QP-4400 (2%)	175.0	21.00
Water	87.5	10.50
Rhoplex AC-34 (46%)	218.0	24.41
Colloid 581-B	1.0	0.13
Ammonia (28%)	2.0	0.27

Visc. (KU)	70
PVC (%)	42
lb/gal	12.03

No. 4

(Premium Quality)

Ethylene Glycol	Premix	25.0
Natrosol 250 MHR		2.8
Water		150.0

Tamol 850 (30%)	10.5
Triton CF-10	2.5
KTPP	1.5
Nopco NXZ	1.0
Propylene Glycol	34.0
Rutile Titanium Dioxide	264.8
Zinc Oxide	50.0
Talc	97.7
Calcium Carbonate	78.7

Grind for 5 min (Cowles 3800–4500 fpm) then add:

Magnesium Aluminum Silicate	10.0

Grind an additional 15 min (Cowles 3800–4500 fpm) then add at lower speed:

Rhoplex AC-64 (60.5%)		390.0
Nopco NXZ		3.0
Texanol	Premix	14.1
Skane M-8	Premix	2.0
Ammonium Hydroxide (28%)		2.0
Water and/or **Natrosol 250 MHR** (2½% sol'n.)		86.0

PVC (%)	40.7
Total Solids by Wt. (%)	60.6
Total Solids by Vol. (%)	42.5
pH–Initial	9.5
Visc.–Initial (KU)	74–78

No. 5

(Economy)

Water		220.0
Tamol 850 (30%)		8.7
KTPP		1.0
Foamaster VL		2.0
Ethylene Glycol	Premix	30.0
Natrosol 250 MHR (100%)	Premix	2.5
Ti Pure R-933		150.0

AZO-11	25.0
Icecap K	50.0
Snowflake	240.6

Grind 5 min, then add:

Attagel 50	5.0

Grind the above for 15 min.
Let down:

Rhoplex AC-64 (60.5%)	266.9
Foamaster VL	1.0
Texanol	8.0
Skane M-8	1.0
Water and/or **Natrosol 250 MHR** (2.5%)	160.3

PVC (%)	52
Total Solids by Vol. (%)	36
pH	9.5
Visc. (KU)	92-98

No. 6

(Economy)

Water		235.0
Tamol 850 (30%)		6.6
KTPP		0.5
Triton CF-10		1.0
Foamaster VL		1.0
Natrosol 250 MHR (100%)	} Premix	2.5
Ethylene Glycol		25.0
Ti Pure R-933		115.0
AZO-11		25.0
Icecap K		116.3
Snowflake		84.9

Grind 5 min, then add:

Attagel 50	5.0

Grind the above for 15 min.

Let down:

Rhoplex AC-64 (60.5%)	173.7
Texanol	5.1
Foamaster VL	3.0
Skane M-8	1.0
Water and/or **Natrosol 250 MHR** (2.5%)	280.8

PVC (%)	55
Total Solids by Vol. (%)	25
pH	9.4
Visc. (KU)	92–98

No. 7

(Sheen)

	lb	gal
Water	54.0	6.48
Tamol 850 (30%)	14.2	1.42
Triton X-100	2.5	0.28
Experimental Stabilizer 612	3.7	0.48
Propylene Glycol	60.0	6.98
Nopco NXZ	1.0	0.13
Rutile Titanium Dioxide	250.0	7.30
Zinc Oxide	25.0	0.54
ASP-170	94.9	4.41

Grind for 15–20 min (Cowles 4000–4400 fpm) and let down at slower speed.

		lb	gal
Rhoplex AC-64		354.3	39.50
Water		15.0	1.80
Texanol	Premix	10.1	1.28
Nopco NXZ	Premix	1.0	0.13
Propylene Glycol	Premix	38.8	4.51
Skane M-8	Premix	2.0	0.23
Natrosol 250 MR (3% sol'n.)		135.0	16.47
Water		69.4	8.33

PVC (%)	35
Total Solids by Wt. (%)	52.5
Total Solids by Vol. (%)	35
pH–Initial	9.4-9.7
Visc.–Initial (KU)	72-76

No. 8

(Sheen)

	lb	gal
Water	54.0	6.48
Tamol 850 (30%)	14.2	1.42
Triton X-100	2.5	0.28
Experimental Stabilizer 612	3.7	0.48
Propylene Glycol	60.0	6.98
Nopco NXZ	1.0	0.13
Titanium Dioxide	250.0	7.30
Zinc Oxide	25.0	0.54
ASP-170	94.9	4.41

Grind on a high-speed dissolver (Cowles) for 15-20 min at 4000-4400 fpm. Let down at slower speed as follows:

Rhoplex AC-707 (65%)	327.1	36.47
Water	15.0	1.80
Texanol	10.1	1.28
Nopco NXZ	1.0	0.13
Propylene Glycol	38.8	4.51
Skane M-8	2.0	0.23
Natrosol 250 MR (3% sol'n.)	140.0	16.80
Water	89.7	10.76

Visc.–Initial (KU)	74-78
pH–Initial	9.5-9.7
PVC (%)	35
Total Solids by Vol. (%)	35

No. 9

(Emulsion)

		lb	gal
Water		200.0	24.00
Nopco NXZ		1.0	1.30
Ethylene Glycol	Premix	27.9	3.00
Hydroxyethyl Cellulose	Premix	2.0	–
Tamol 850 (30%)		10.5	1.05
KTPP		1.5	–
Triton CF-10		2.5	0.28
Rutile Titanium Dioxide		250.0	7.60
Zinc Oxide		25.0	0.54
Talc		175.0	7.35

Grind the above materials in a high-speed mill (Cowles, 3800-4200 fpm for 10-15 min) then add the following at slower speed:

	lb	gal
Rhoplex AC-388 (50%)	440.0	50.00
Nopco NXZ	1.0	0.13
Texanol	11.0	1.39
Skane M-8	0.5	0.06
Water and/or HEC (2.5% sol'n.)	37.2	4.47

PVC (%)	39.7
Total Solids by Vol. (%)	39.0
Total Solids by Wt. (%)	56.5
Visc. (KU)	77
pH	9.3-9.5

No. 10

	lb	gal
Hydroxyethyl Cellulose (2.5%)	85.0	10.30
Water	62.5	7.50
Tamol 850 (30%)	10.5	1.05
Triton CF-10	2.5	0.28
KTPP	1.5	0.07
Nopco NXZ	1.0	0.13

Ethylene Glycol	25.0	2.69
Titanium Dioxide	237.5	7.22
Zinc Oxide	50.0	1.07
Talc	187.7	8.11

Grind the above materials in a high-speed mill (Cowles, 3800–4500 fpm for 20 min) then add the following at slower speed:

Rhoplex AC-388 (50%)		459.8	52.23
Nopco NXZ		1.0	0.13
Tributyl Phosphate	Premix	11.0	1.36
Propylene Glycol	Premix	34.0	3.95
Skane M-8 (45%)	Premix	2.0	0.23
Ammonium Hydroxide (28%)		1.0	0.13
Water		29.6	3.55

PVC (%)	40.0
Total Solids by Vol. (%)	41.0
Total Solids by Wt. (%)	58.7
Visc. (KU)	70–75
pH	9.5

No. 11

(Flat, Alkyd Modified)

Charge to tank in order shown. Add at low speed.

	lb	gal
Water	125.0	15.00
Dowicil 75	2.0	0.17
Tamol 850	12.0	1.25
KTPP	2.0	0.20
Ammonium Hydroxide (28%)	0.2	0.03
Igepal CO-630	3.0	0.38
AZO-ZZZ-33	50.0	1.07
Strodex PK-90	2.0	0.22
Ethylene Glycol	14.0	1.50
Ti Pure R-900	220.0	6.65
Minex 4	115.0	5.34

Shift to high speed. Grind for 5–10 min. Proceed.

Mica P-80-f		30.0	1.39
Drew Y-200		1.5	0.19
Ethylene Glycol	Premix*	16.0	1.75
Zanflo		4.5	0.48
Water		117.0	14.00

Premix **Zanflo**, add to batch slowly at high speed. Grind at high speed for 20 min or to a 7 Hegman grind.

Rhoplex AC-388		315.0	36.00
Drew Y-200	Premix**	0.5	0.06
Texanol		9.0	1.13
Alkyd		73.8	8.38
Water and/or **Zanflo** (1½% sol'n.)		58.0	7.00

Drop grind paste into let down portion, use slow speed to mix. Mix well before continuing. Adjust viscosity as needed.

---

* Add **Zanflo** to ethylene glycol, mix, immediately add to tank at high speed. **Zanflo** may be slowly sifted in dry or added as a solution.

** Premix if convenient. If not, drop grind into latex and proceed.

		No. 12 *(Economy, Unmodified)*		No. 13 *(Economy, (Alkyd Modified)*	
		lb	gal	lb	gal
Hydroxyethyl Cellulose	Premix	2.5	0.22	2.5	0.22
Ethylene Glycol		18.0	1.94	18.0	1.94
Water		150.0	18.00	150.0	18.00
Tamol 850 (30%)		12.3	1.23	11.7	1.17
KTPP		1.5	0.07	1.5	0.07
Triton CF-10		1.0	0.11	1.0	0.11
Antifoamer		1.0	0.13	1.0	0.13
Titanium Dioxide		150.0	4.56	150.0	4.56
Zinc Oxide		25.0	0.54	25.0	0.54
Calcium Carbonate		200.0	8.86	200.0	8.86
Talc		110.2	4.76	98.2	4.24

Grind the above materials on a high-speed mill (Cowles, 3800–4500 fpm for 10–15 min) and let down, at a slower speed, as follows:

Rhoplex AC-388 (50%)	323.0	36.69	276.2	31.38
Long Oil Alkyd (drier treated*)	–	–	16.9	2.02
Coalescent } Premix	7.9	1.00	9.5	1.20
Antifoamer } Premix	1.0	0.13	1.0	0.13
Skane M-8 } Premix	1.0	0.12	1.0	0.12
Water	139.7	16.76	149.9	17.99
Hydroxyethyl Cellulose (2.5% sol'n.)	40.0	4.88	60.0	7.32

* Drier is 0.5% of 6% cobalt naphthenate, 0.5% of 6% manganese naphthanate, and 5.6% of 6% zirconium octoate.

PVC (%)	52.0	52.0
Total Solids by Vol. (%)	36.0	35.0
Visc.–Initial (KU)	76–78	78–80
pH	9.3–9.5	9.3–9.5

Acrylic Latex White Exterior Paint

(Gloss)

	lb	gal
Water	60.0	7.2
KTPP	2.0	0.1
Igepal CO-630	4.0	0.5
Balab 748	3.0	0.4
Cosan PMA-30	2.0	0.2
Ethylene Glycol	20.0	2.1
Titanox 2020	210.0	6.1
Butyl **Carbitol**	16.0	2.0

Disperse in high-speed mill; then add the following ingredients:

Water	23.0	2.7
Strodex MO-100	2.0	0.2
N,N-Dimethylethanolamine (DMEA)	1.0	0.1

Ucar Vehicle 4358	569.0	64.9
Balab 748	1.0	0.1
Ammonium Hydroxide (28% aq. sol'n.)	4.0	0.4
Acrysol G-110 (11% NV)	32.0	3.8
Cargill 7407 (80% NV)	48.0	5.5
Cobalt Drier (6%, W.D. type)	1.0	0.1
Zirconium Drier (6%)	1.0	0.1
Cosan PMA-30	2.0	0.2
Dibutyl Phthalate	7.0	0.8
Water	21.0	2.5

PVC (%)	15.8
Total Solids by Wt. (%)	51.1
Total Solids by Vol. (%)	42.5
Visc. (s—#4 Ford cup)	20–25
pH	8.3–8.7

Exterior White Epoxy Maintenance Paint

		lb	gal
A	Rutile Titanium Dioxide	331.47	9.46
	EKS-2002 (75% NV)	426.44	46.77
	Beetle 216-8 (60% NV)	16.11	1.89
	Methyl Ethyl Ketone	95.72	14.26
	Cellosolve Solvent	47.39	6.11
	Butyl **Cellosolve**	11.36	1.51
B	**Polyamide 325**	106.14	13.11
	Xylene	26.82	3.72
	Cellosolve Solvent	24.49	3.17

Procedure:

Pebble mill grind ingredients in Part A until a good dispersion is obtained. The hardener portion, Part B, can be mixed on any standard mixing equipment. Do not mix the two components until ready to apply.

Mixing Ratio:

Resin Portion by Wt. (lb)	5.9
Hardener Portion by Wt. (lb)	1.0

After the two components are mixed, a 30-min induction period is recommended before application of the coating.

PVC (%)	30.52

Exterior Latex White Paint

Formula No. 1

(Semigloss)

	lb	gal
Charge the following to the tank:		
Tamol 731 (25%)	11.0	1.20
Nopco NDW	2.0	0.27
Propylene Glycol	67.0	7.72
Add the following at low speed, 1900 fpm:		
Rutile Titanium Dioxide	275.0	7.87

Increase speed to 5400 fpm and grind for 20 min; then add the following at low speed, 1900 fpm:

	lb	gal
Propylene Glycol	113.0	13.11
Rhoplex AC-22 (44.5%)	440.0	49.72
Rhoplex AC-61 (46%)	106.0	11.98
Nopco NDW	2.0	0.27
Preservative } Premix	1.0	0.12
Water } Premix	62.3	7.50
Triton GR-7 (64%)	2.0	0.23
Hydroxyethyl Cellulose (2.5% sol'n.)	46.6	5.64

Total Solids (%)	46.3
PVC (%)	23.5
Visc–Initial (KU)	68

No. 2

(Semigloss)

	lb	gal
Charge the following to the tank:		
Tamol 731 (25%)	7.2	0.78
Balab 748	2.1	0.28
Propylene Glycol	51.2	5.94
Add the following at low speed, 1900 fpm:*		
Rutile Titanium Dioxide	235.7	6.74
Increase speed to 5400 fpm and grind for 20 min; then add the following at low speed, 1900 fpm:		
Rhoplex AC-22 (44.5%)	522.6	59.02
Rhoplex AC-73 (46%)	111.7	12.62
Preservative } Premix	1.0	0.12
Balab 748 } Premix	2.1	0.28
Butyl **Cellosolve** } Premix	6.2	0.82
Acrysol G-110 (22%)	5.1	0.58
Water	102.5	12.30
Triton GR-5 (60%)**	2.0	0.27
Total Solids (%)	49.6	
PVC (%)	18.8	
Visc.–Initial (KU)	65	

* This formulation is made in a high-speed disc-type disperser such as a Cowles.

** A lower foaming wetting agent such as **Triton GR-7** can be substituted, if desired.

No. 3

(Semigloss, Extender Modified)

	lb	gal
Charge the following to the tank:		
Tamol 731 (25%)	10.2	1.12
Nopco NDW	1.9	0.25
Propylene Glycol	62.1	7.20

Add the following at low speed:

Rutile Titanium Dioxide (high gloss)	254.9	7.43
Barytes	46.3	1.25

Grind the above in a high-speed mill (Cowles, 3800–4500 fpm for 10–15 min) and let down at lower speed with the following:

Propylene Glycol		104.7	12.13
Rhoplex AC-490 (46.5%)		487.5	55.09
Nopco NDW		1.9	0.25
Super Ad-It	Premix	0.9	0.11
Water	Premix	56.7	6.88
Butyl **Cellosolve**	Premix	22.6	2.88
Triton GR-7		1.9	0.21
Hydroxyethyl Cellulose (2.5% sol'n.)		43.0	5.20

PVC (%)	27.1
Total Solids by Vol. (%)	32.0
Total Solids by Wt. (%)	48.2
Visc.—Initial (KU)	73

No. 4

(Semigloss, High-Hiding)

	lg	gal
Charge the following to the tank:		
Tamol 731 (25%)	9.0	0.98
Balab 748	2.0	0.26
Propylene Glycol	70.0	8.12

Add the following at low speed, 1900 fpm:

Rutile Titanium Dioxide	275.0	7.87
Calcium Carbonate	16.0	0.71

Increase speed to 5400 fpm and grind for 20 min; then add the following at low speed, 1900 fpm:

Propylene Glycol	30.0	3.48
Ethylene Glycol	40.0	4.30
Rhoplex AC-22 (44.5%)	550.0	62.15

Pine Oil	Premix	1.8	0.20
Preservative		0.9	0.11
Balab 748		1.5	0.20
Water	Premix	85.8	10.30
Acrysol G-110 (22%)		7.5	0.89
Triton GR-5 (60%)		2.2	0.24

Total Solids	49.5
PVC (%)	25
Visc.–Initial (KU)	65

No. 5

(Semigloss, Economy)

		lg	gal
Charge the following to the tank:			
Tamol 731 (25%)*		7.06	0.77
Balab 748		1.80	0.25
Propylene Glycol		60.00	6.96
Add the following at low speed, 1900 fpm:			
Rutile Titanium Dioxide		220.75	6.31
Calcium Carbonate		17.66	0.79
Increase speed to 5400 fpm and grind for 20 min; then add the following at low speed, 1900 fpm:			
Rhoplex AC-22 (44.5%)		549.22	62.07
Propylene Glycol		45.96	4.59
Ethylene Glycol**	Premix, then add with good agitation	35.32	3.78
Pine Oil		1.77	0.19
Preservative		0.88	0.11
Balab 748		1.70	0.23

Acrysol G-110 (22%) }	Premix then add with	8.83	0.99
Water }	good agitation	105.96	12.72
Triton GR-5 (60%)†		2.21	0.26

Total Solids (%)	45
PVC (%)	22.4
Visc.–Initial (KU)	63–65

---

* **Tamol 731** (25%) must be incorporated before pigment is added.
** Ethylene glycol can be used for dilution purposes.
† A lower foaming wetting agent such as **Triton GR-7** can be substituted, if desired.

No. 6

(High Gloss)

	lb	gal
Charge the following to the tank in the order listed and mix until uniform:		
Ethylene Glycol	15.0	1.61
Nopcosant K	4.2	0.38
KTPP	0.4	0.20
Next add the following:		
Propylene Glycol	95.0	11.05
Foamaster G	4.2	0.55
At low speed (1900 fpm) on a Cowles type mixer add:		
Ti Pure R-900 or **Titanox CL-NC** or **Unitane OR-580**	275.0	8.20
Increase speed to 5400 fpm and grind for 20 min; then add the following at low speed (1900 fpm):		
Igepal CO-610	4.4	0.50
BEP (1-butoxyethoxy-2-propanol)	10.0	1.30
Ammonium Hydroxide (28%)	4.0	0.51
Wallpol 40-138	590.0	65.50
Foamaster G	4.2	0.55
Ammonium Hydroxide (28%)	8.3	1.01
Water	70.0	8.42

Total Solids (%)	53.1
Total NV by Vol. (%)	38.7
PVC (%)	20.3
pH	9.0-9.3
Visc. (KU)	80-85

No. 7

(Alkyd Modified)

	lb	gal
Dispersion:		
Water (as needed)	361	43.3
Biocide	5	0.3
Nopco NXZ	2	0.3
Ethylene Glycol	28	3.0
Natrosol 250 HR	4	0.3
Strodex MOK-70	5	0.5
Tamol 850	5	0.5
Titanox 2030	225	7.0
AZO-33	25	0.5
Asbestine 3X	150	6.5
Grind at high speed.		
Let down:		
Igepal CO-630	5	0.6
Texanol	8	1.0
Polyco 2355	273	30.0
Premix and add:		
Aroplaz 1271	50	6.0
Zirco Catalyst (6%)	1	0.1
Cobalt Naphthenate (6%)	1	0.1

Total Vehicle Solids by Vol. (gal)	21.0
Total Solids by Vol. (gal)	35.0
Total Solids by Wt. (%)	53.0
PVC (%)	40.0

No. 8

(Alkyd Modified)

Water	328.0
KTPP	2.0
Daxad 30	9.0
Igepal CO-630	3.0
Advacide 60	1.5
Defoamer	2.0
Celite 281	25.0
Nytal 300	75.0
Titanox RA-NC	245.0
Titanox AC-G	30.0
EP 303 Lead Silicate	133.0
T&W 345-100 Super Alkyd	62.0
Lead Drier (24%)	1.0
Cobalt Drier (6%)	0.4
Ethylene Glycol	25.0
Carbitol Acetate	9.0
Natrosol 250 MR	4.0
Parco 37-C-55	310.0

PVC (%)	39.6
Total Solids (%)	58.7
Visc. (KU)	84-90
lb/gal	12.3

No. 9

Water	200
Tamol 731	11
Triton CF-10	2
Nopco NXZ	1
Ethylene Glycol	25
PMA-60	2
Natrosol 250 MR	3
Titanox CL-NC	240
Ti Pure A-168-LO	10
Asbestine 325	100

Snowflake	100
ParCryl 300	495
Nopco NXZ	1
Pine Oil	6
Ammonium Hydroxide (28%)	2

PVC (%)	40.2
Total Solids (%)	58.1
Visc. (KU)	85-89
pH	9.3-9.9

Polyvinyl Acetate Exterior White Paint with Triamino Ester

Formula No. 1

	lb	gal
Premix:		
Water	150.0	18.00
Strodex MOK-70	4.0	0.44
Triton X-100	4.0	0.45
KTPP	1.0	–
Ammonia (28%)	4.0	0.52
Increase speed and add slowly:		
Chemacoil TA-100	150.0	19.23
Adjust speed to disperse pigments:		
Azo ZZZ-33	50.0	1.07
Ti Pure R-901	225.0	6.57
Titanox A-168-LO	25.0	0.77
Duramite	150.0	6.69
Mica (325 mesh, water-ground)	12.5	0.53
Ethylene Glycol	18.5	2.00
Nuodex PMA-18	3.0	0.26
Colloid 581-B	2.0	0.30

Slow to mixing speed:

Cellosize QP-15000 (2%)	119.0	14.28
Everflex BG (52%)	282.0	31.68
Super-Cobalt Drier	1.0	0.13
Visc.—Initial (KU)	87	
PVC (%)	31.4	

No. 2

	lb	gal
Premix:		
Water	167.0	20.00
AMP	7.0	0.91
Triton X-100	4.0	0.45
Increase speed and add slowly:		
Chemacoil TA-100	150.0	19.23
Disperse pigments:		
Ti Pure FF	25.0	0.75
Ti Pure R-901	225.0	6.44
Mica (325 mesh, water-ground)	12.5	0.58
Celite 281	25.0	1.30
Nytal 300	150.0	6.64
Decrease speed and add slowly:		
Colloid 581-B	2.0	0.25
Ethylene Glycol	18.6	2.00
Nuodex PMA-18	3.0	0.30
Everflex BG (52%)	282.0	31.69
Super-Cobalt Drier	1.5	0.19
Cellosize QP-15000 (2%)	101.0	12.02
Visc. (KU)	77	
pH	8.8	
PVC (%)	32	
Total Solids by Vol. (%)	63.4	

No. 3

	lb	gal
Premix:		
Water	170.0	20.40
Igepal CTA-639	12.0	1.36
KTPP	1.0	–
Victawet 35-B	1.0	0.11
Increase speed and add slowly:		
Chemacoil TA-100	150.0	19.23
Adjust speed to disperse pigments:		
Titanox RCHT-X	375.0	13.84
Celite 281	50.0	2.60
Mica (325 mesh, water-ground)	25.0	1.05
Ethylene Glycol	18.5	2.00
Nuodex PMA-18	3.0	0.26
Colloid 581-B	2.0	0.30
Slow to mixing speed:		
Cellosize QP-15000 (2%)	98.0	11.76
Everflex BG Emulsion (52%)	266.0	29.86
Super-Cobalt Drier	1.0	0.13
Visc.—Initial (KU)	78	
PVC (%)	33.5	

No. 4

	lb	gal
Premix:		
Water	133.3	16.00
Daxad 30	8.4	0.98
Igepal CTA-639	2.0	0.23
KTPP	1.0	–

Increase speed and add slowly:

Chemacoil TA-100	94.0	12.05
Adjust speed to disperse pigments:		
Ti Pure R-901	225.0	6.57
Titanox A-168-LO	25.0	0.77
Nytal 300	50.0	2.10
Mica (325 mesh, water-ground)	30.0	1.26
Polyglycol P-1200	2.0	0.24
Colloid 581-B	1.0	0.15
Ethylene Glycol	20.0	2.16
Nuodex PMA-18	5.0	0.45
Slow to mixing speed:		
Cellosize QP-4400 (3%)	140.0	16.80
Lead Naphthenate (24%)	2.0	0.21
Cobalt Naphthenate (6%)	0.5	0.06
Carbitol Acetate	20.0	2.36
Everflex BG Emulsion (52%)	425.0	47.73
Visc.—Initial (KU)	87	
PVC (%)	32	

No. 5

	lb	gal
Premix:		
Water	150.0	18.00
Triton X-100	5.0	0.56
Daxad 30	10.0	1.18
Victawet 35-B	3.0	0.31
Increase speed and add slowly:		
Chemacoil TA-100	177.0	22.67
Adjust speed to disperse pigments:		
Azo ZZZ-55-LO	125.0	2.67
Ti Pure R-901	175.0	5.11
Titanox A-168-LO	50.0	1.54

Nytal 300	50.0	2.10
Ethylene Glycol	40.0	4.32
Super-Ad-It	2.0	0.25
Water	61.0	7.32
Colloid 581-B	2.0	0.30

Slow to mixing speed:

Natrosol 250 HR (2½%)	193.0	23.16
Everflex BG (52%)	114.0	12.80
Super-Cobalt Drier	1.0	0.13

Visc.–Initial (KU)	92
PVC (%)	26.7

No. 6

	lb	gal
Premix:		
Water	175.0	18.00
Daxad 30	8.0	0.94
Victawet 35-B	2.5	0.26
Increase speed and add slowly:		
Chemacoil TA-100	150.0	19.23
Adjust speed to disperse pigments:		
Ti Pure R-901	225.0	6.57
Titanox A-168-LO	25.0	0.77
Asbestine 3X Talc	150.0	6.31
Celite 281	25.0	1.30
Mica (325 mesh, water-ground)	12.5	0.53
Ethylene Glycol	18.5	2.00
Nuodex PMA-18	3.0	0.26
Colloid 581-B	2.0	0.30
Slow to mixing speed:		
Cellosize QP-15000 (2%)	105.0	12.60
Aircoflex 510 (55%)	262.0	28.79

Ammonia (28%)	4.0	0.52
Super-Cobalt Drier	1.0	0.13
Visc.–Initial (KU)	82	
PVC (%)	32.3	

No. 7

	lb	gal
Premix:		
Water	142.0	17.03
Strodex PK-90	2.0	0.21
Triton X-100	6.0	0.68
Ammonia (28%)	4.0	0.52
Increase speed and add slowly:		
Chemacoil TA-100	125.0	16.01
Adjust speed to disperse pigments:		
Titanox RA-47	175.0	5.11
Duramite	175.0	7.81
Celite 281	50.0	2.60
Hexylene Glycol	23.0	2.92
Super-Ad-It	7.0	0.86
Water	50.0	6.00
Colloid 581-B	2.0	0.30
Slow to mixing speed:		
Natrosol 250 HR (2%)	150.0	18.00
Ucar 180 (55%)	227.0	24.90
Super-Cobalt Drier	1.3	0.16
Visc.–Initial (KU)	82	
PVC (%)	35.0	

No. 8

(Economy)

	lb	gal
Premix:		
Water	125.0	15.00
Daxad 30	8.0	0.83
Tergitol NPX	2.5	0.28
Increase speed and add slowly:		
Chemacoil TA-100	101.4	13.00
Disperse pigments:		
Rutile Titanium Dioxide	150.0	4.29
Anatase Titanium Dioxide	25.0	0.77
Mica (325 mesh, water-ground)	25.0	1.16
Whiting	100.0	4.46
Slow to mixing speed:		
Nuodex PMA-18	1.0	0.10
Water	46.0	5.50
Colloid 581-B	2.0	0.25
Cellosize QP-15000 (2%)	325.0	39.00
Super-Cobalt Drier	0.5	0.07
Polyvinyl Acetate Copolymer Resin Latex (55%)	163.5	18.00

Visc.–Initial (KU)	86
PVC (%)	32.4
Total Solids by Vol. (%)	33
Total Solids by Wt. (%)	46.8
lb/gal	10.47

Vinyl Acrylic Exterior White Paint

Formula No. 1

(Nonmercurial, Mildew-Resistant)

		lb	gal
Titanox CL-NC		220.0	6.4
Natrosol 250 HR (1½% sol'n.)		130.0	15.5
Snowflake		120.0	5.3
Water		100.0	17.0
Nytal 300		55.0	2.3
Ethylene Glycol		25.0	2.6
Tamol 731		4.0	0.4
Igepal CTA-639		2.0	0.2
Colloid 677		1.5	0.2
KTPP		1.0	0.1
Metasol TK-100		1.0	0.1
Merbac-35		1.0	0.1
Disperse in high-speed mill, then add slowly with agitation:			
Resyn 2345		290.0	32.0
Natrosol 250 HR (1½% sol'n.)		100.0	11.9
Water		42.0	5.0
Colloid 677		1.0	0.1
Castung 235	Premix, then add	40.0	5.0
Igepal CTA-639	Premix, then add	5.0	0.6
Advacar (24% lead drier)	Premix, then add	2.0	0.2
Advacar (6% cobalt drier)	Premix, then add	1.0	0.1

lb/gal	11.4
PVC (%)	29.6
Total Solids (%)	53.5
Visc. (KU)	75–80

No. 2

(Alkyd-Modified)

Water	121.0
Tamol 731	15.0
Triton CF-10	2.5

Nopco NXZ	1.0
Metasol 57	1.8
Titanox CL-NC	300.0

Mix on a high-speed mill and add:

Texanol	Premix	6.0
Natrosol 250 MR	Premix	3.5
Ethylene Glycol	Premix	25.0
T&W 345-100 Super Alkyd		30.0
Lead (24%)		1.0
Manganese (6%)		0.5
ParCryl 400		400.0
Nopco NXZ		1.0
Ammonium Hydroxide		2.0
Propylene Glycol		35.0
Water		17.0

PVC (%)	40
Total Solids (%)	56
Visc. (KU)	73–77
pH	9.5

No. 3

(For New Wood and Chalky Surfaces)

	lb	gal
Water	115.0	13.80
Methocel 4000 cps (90 HG-DGS)	3.0	0.26
Potassium Carbonate	1.0	0.05
KTPP	1.0	0.05
Tamol 731 (25%)	3.0	0.33
Emulphor EL-719	3.0	0.34
R&R 551	4.0	0.46
Rutile Titanium Dioxide (maximum chalk resistance)	225.0	6.45
Anatase Titanium Dioxide (medium oil absorption)	25.0	0.77
Asbestine 3X	75.0	3.16

Mica (325 mesh, water-ground)	25.0	1.07
Carbitol Acetate	4.0	0.48
Ethylene Glycol	18.0	1.94
Troysan PMA-30	10.0	1.02
Nopco NXZ	2.0	0.27

Disperse in high-speed mill, then add slowly with agitation:

Water	100.0	12.00
Polyco 2155	392.0	42.90
Water and/or **Methocel** (4000 cps, 3% sol'n.)	125.0	15.00

Note:

Add 8 oz of **Castung 235** composition* per gal.

---

* Composition of **Castung 235**

Castung 235	100.0
Lead Naphthenate	3.8
Cobalt Naphthenate	1.7
PMD-11	2.5

lb/gal	1128
PVC (%)	33.9
Total Solids (%)	51.6
Visc. (KU)	80–85
pH	8.0–8.5

Vinyl Chloride-Acrylic Exterior White House Paint

Formula No. 1

	lb	gal
Water	60.00	7.19
Triton X-100	3.00	0.34
Ethylene Glycol	20.00	2.14
Amino Methyl Propanol	5.50	0.70
Tamol 731	9.00	0.98
Hercules 357	5.00	0.66

Blend above and add in order listed

Titanox A-LO	25.00	0.77
Nytal 300	140.00	5.96
Titanox RA-50	245.00	7.00

Mix until smooth (Cowles 2000–3000 rpm for 10–15 min)

Water	15.00	1.80
KTPP	2.00	0.10

Mix 5 min on Cowles

Natrosol 250 MR (2½%)	100.00	11.90
Long-oil Alkyd Intermediate, BFG-12273	96.00	11.58

Mix 5 min on Cowles
Wash and reducer

Water	60.00	7.19
Paste	785.50	58.31

Add under a slow-speed agitation:

Geon 450X20 Latex	260.00	28.26
Hercules 357	1.00	0.13
Triton X-100	5.00	0.56
Water or **Natrosol 250 MR** (2½%)	106.00	12.71

PVC (%)	35
Visc. (KU)	75–80
pH	9.1–9.5

(If pH is low, adjust with amino methyl propanol.)

No. 2

(Containing Zinc Oxide)

	lb	gal
Water	105.7	12.6
KTPP	11.3	–
Triton X-100	11.3	1.2
Tamol 731	12.7	1.4
Hercules 357	7.0	0.9

Perma Kleer 80	5.0	0.4
Zinc Oxide XX601	175.0	3.7
Nytal 300	200.0	8.4
Titanox RA-NC	200.0	5.7

Mix until smooth on high-shear dispersion equipment (Cowles or equal) at 2000–3000 rpm for 15–20 min.

Ethylene Glycol	28.2	3.0
Cellosize WP-4400 (2½%)	141.0	16.9

Premix following with slow agitation until homogenous, and mix with above pigment paste at 300–1000 rpm.

BFG-12274	112.8	13.6
Geon 450X20 Latex	305.5	32.4
Hercules 357	1.4	0.2

lb/gal	13.11
Total NV by Wt. (%)	65.51
Total NV by Vol. (%)	45.80
PVC (%)	38.85

Note:

Accelerated stability tests project a stability of at least one year for this product.

Acrylic Exterior Deep Tone Paint

Formula No. 1

(Barn Red)

Tamol 731 (25%)	5.8
Triton CF-10	2.0
Foamaster VL	1.0
Ethylene Glycol	35.0
Natrosol 250 MHR (2.5%)	50.0
Water	125.0
RO-8097	80.0
RO-2899	20.0
Silica	305.7

Grind 5 min at high speed, then add:		
Attagel 50		10.0
Grind 10–15 additional min and let down:		
Rhoplex AC-64 (60.5%)		389.3
Foamaster VL		3.0
Texanol		3.0
Super Ad-It		9.0
Natrosol 250 MHR (2.5%) and/or Water		100.5
PVC (%)	40.0	
Total Solids by Vol. (%)	42.0	
pH	9.2	
Visc. (KU)	82–86	

No. 2

(Gold Yellow)

Tamol 731 (25%)		9.3
Triton CF-10		2.0
Foamaster VL		1.0
Ethylene Glycol		35.0
Natrosol 250 MHR (2.5%)		50.0
Water		125.0
YLO-1788		100.0
Silica		293.8
Grind 5 min at high speed, then add:		
Attagel 50		10.0
Grind 10–15 additional min and let down:		
Rhoplex AC-64 (60.5%)		389.3
Foamaster VL		3.0
Texanol		3.0
Super Ad-It		9.0
Natrosol 250 MHR (2.5%) and/or Water		97.2
PVC (%)	40.0	
Total Solids by Vol. (%)	42.0	
pH	9.1	
Visc. (KU)	82–86	

No. 3

(Green)

Tamol 731 (25%)	6.3
Triton CF-10	2.0
Foamaster VL	1.0
Ethylene Glycol	35.0
Natrosol 250 MHR (2.5%)	50.0
Water	125.0
X-1134	100.0
Silica	307.0

Grind for 5 min at high speed, then add:

Attagel 50	10.0

Grind for 10–15 additional min and let down:

Rhoplex AC-64 (60.5%)	389.3
Foamaster VL	3.0
Texanol	3.0
Super Ad-It	9.0
Natrosol 250 MHR (2.5%) and/or Water	100.5

PVC (%)	39.9
Total Solids by Vol. (%)	42.0
pH	9.1
Visc. (KU)	82-86

	No. 4 *(Dry Ground, Dark Brown)*		No. 5 *(Dry Ground, Dark Brown, Alkyd Modified)*	
	lb	gal	lb	gal
Tamol 731 (25%)	9.8	1.06	9.8	1.06
Triton CF-10	2.0	0.23	2.0	0.23
Antifoamer	1.0	0.13	1.0	0.13
Ethylene Glycol	35.0	3.77	35.0	3.77

Hydroxyethyl Cellulose (2.5% sol'n.)	100.0	12.12	100.0	12.12
Water	60.0	7.20	60.0	7.20
Preservative (57% mercury)	1.8	0.11	1.8	0.11
Mapico 421	75.0	1.89	75.0	1.89
Mapico 418	75.0	1.89	75.0	1.89
Silica	286.8	13.02	286.8	13.02

Grind the above materials in a high-speed mill (Cowles, 3800-4500 fpm for 10-15 min) and let down, at a slower speed, as follows:

Rhoplex AC-388 (50%)	471.0	53.51	400.4	45.49
Long Oil Alkyd*	–	–	31.5	3.78
Antifoamer (Premix)	1.0	0.13	1.0	0.13
Tributyl Phosphate (Premix)	10.0	1.23	10.0	1.23
Ammonium Hydroxide (28%) (Premix)	2.0	0.27	2.0	0.27
Water and/or Hydroxyethyl Cellulose (2.5%)	28.7	3.44	64.0	7.68

* Drier treated with 0.5% of 6% cobalt, 0.5% of 6% manganese, and 1.4% of 24% lead prior to incorporation of alkyd.

PVC (%)	40.0	40.0
Total Solids by Vol (%)	42.0	42.0
Total Solids by Wt. (%)	58.0	57.9
pH	≈9.4-9.7	≈9.4-9.7
Visc.–Initial (KU)	72-78	72-78

No. 6

(Dark Brown)

Tamol 731 (25%)	9.8
Triton CF-10	2.0
Foamaster VL	1.0
Ethylene Glycol	35.0
Natrosol 250 MHR (2½%)	50.0
Water	125.0

Mapico 421	75.0
Mapico 418	75.0
Silica	275.8

Grind for 5 min at high speed, then add:

Attagel 50	10.0

Grind 10–15 additional min and let down:

Rhoplex AC-64 (60.5%)	389.3
Foamaster VL	3.0
Texanol	3.0
Super Ad-It	9.0
Natrosol 250 MHR (2½%) and/or Water	96.9

PVC (%)	40.0
Total Solids by Vol. (%)	42.0
pH	9.2
Visc. (KU)	82-86

No. 7

(Black)

Tamol 731 (25%)	6.3
Triton CF-10	2.0
Foamaster VL	1.0
Ethylene Glycol	35.0
Natrosol 250 MHR (2.5%)	50.0
Water	125.0
Germantown Lamp Black	20.0
Silica	330.4

Grind 5 min at high speed, then add:

Attagel 50	10.0

Grind 10–15 additional min and let down:

Rhoplex AC-64 (60.5%)	389.3
Foamaster VL	3.0

Texanol	3.0
Super Ad-It	9.0
Natrosol 250 MHR (2.5%) and/or Water	100.1

PVC (%)	40.0
Total Solids by Vol. (%)	42.0
pH	9.2
Visc. (KU)	82-86

Latex Exterior White and Light Tint

Formula No. 1

		lb	gal
Charge into mixer under agitation:			
Water		80.0	9.6
KTPP		1.0	0.1
Strodex PK-90		2.0	0.2
Colloid 677		1.5	0.2
Ethylene Glycol		30.0	3.2
Igepal CTA-639		2.0	0.2
Merbac 35		0.5	–
Cellosize QP-4400 (1½% sol'n.)		50.0	6.0
Duramite		75.0	3.3
Metasol TK100		1.0	0.1
Titanox 2061		250.0	7.3
Nytal 300		50.0	2.2
Disperse in high-speed mill, then add slowly with agitation:			
Ucar Latex 2345 or **366**		290.0	31.9
Igepal CTA-639		2.0	0.2
Cargill 7700	Premix	53.5	6.3
Advacar Cobalt Drier (6%)	Premix	3.0	0.3
Advacar Zirco Lead Drier (6%)	Premix	1.5	0.2
Cellosize QP-4400 (1½% sol'n.)		237.0	28.3
Colloid 677		2.0	0.2
Water		32.0	3.9
Texanol		10.0	1.3

PVC (%)	36
lb/gal	11.1
Visc. (KU)	85-90
Total Solids (%)	51.5

No. 2

(Cal Ink System)

		lg	gal
Charge into mixer under agitation:			
Water		227.0	27.2
Merbal 35		0.5	–
Ethylene Glycol		30.0	3.3
Cellosize QP-4400		4.0	0.4
Mix until smooth, then add:			
Colloid 677		1.0	0.1
Strodex MOK-70		5.5	0.7
Tamol 731		8;0	0.9
Igepal CTA-639		2.0	0.2
Optiwite		75.0	4.2
Snowflake		50.0	2.2
Zinc Oxide **EP-417W**		75.0	1.6
Metasol TK100		1.0	0.1
Ti Pure R-902		225.0	6.6
Disperse in high-speed mill, then add slowly with agitation:			
Ucar Latex 2345 or **366**		300.0	33.0
Igepal CTA-639		6.0	0.7
Advacar Cobalt Drier (6%)	Premix, then add slowly	3.0	0.3
Advacar Zirco Lead Drier (6%)		1.5	0.2
Aroplaz 1271		42.0	5.0
Colloid 677		1.5	0.2
Texanol		12.0	1.5
Water		96.0	11.5

PVC (%)	40
lb/gal	11.7
Visc. (KU)	80-85
Total Solids (%)	55.8

Vinyl Ethylene Exterior White and Light Tint

(Alkyd Modified)

lb/100 gal

High-speed mill dispersion in order of addition:

Water	245.0
Merbac-35	0.5
Ethylene Glycol	30.0
Natrosol 250 HR	4.0

Mix at high speed until cellulose is completely hydrated. Reduce speed and add:

Colloid 677	1.0
Igepal CTA-639	2.0
Tamol 850	8.0
Strodex MOK-70	6.0

Increase to high speed and add:

Zopaque R-88	225.0
Snowflake	50.0
Zinc Oxide (N.J. xx-503)	75.0
Kaopaque 30	75.0
Metasol TK-100	1.0

Disperse for 10–15 min, then add:

Airflex 720		300.0
Igepal CTA-639		6.0
Advacar Cobalt Drier (6%)	Premix before adding	3.0
Advacar Zirco Lead Drier (6%)		1.5
Aroplaz 1271		42.0

Colloid 677	1.5
Texanol	12.0
Water and/or **Natrosol 250 HR** (2%)	80.0

PVC (%)	40.0
lb/gal	11.7
Visc. (KU)	85.90
Total NV by Wt. (%)	55.8

Acrylic Exterior Light Tint Base For Glycol Dispersed Colorants

Formula No. 1

Natrosol 250 MHR (100%)	Premix	3.6
Ethylene Glycol		20.0
Tamol 850 (30%)		10.5
KTPP		1.5
Nopco NXZ		1.0
Triton CF-10		2.5
Water		157.5
Ti Pure R-902		170.0
AZO-11		50.0
Silver Bond B		208.6

Grind 5 min at high speed, then add:

Attagel 50	10.0

Grind 10–15 additional min and let down:

Rhoplex AC-64 (60.5%)		370.4
Nopco NXZ		2.0
Tributyl Phosphate		13.4
Propylene Glycol	Premix	48.0
Skane M-8		2.0
Natrosol 250 MHR (2½%) and/or Water		100.5

PVC (%)	40.0
Total Solids by Vol. (%)	40.0
pH	9.2–9.4
Visc. (KU)	88–92

	No. 2		No. 3	
	(Unmodified)		*(Alkyd Modified)*	
	lb	gal	lb	gal
Hydroxyethyl Cellulose				
(2.5% sol'n.)	85.0	10.30	85.0	10.30
Water	68.5	8.22	68.5	8.22
Tamol 731 (25%)	14.1	1.53	14.1	1.53
Triton CF-10	2.5	0.28	2.5	0.28
Antifoamer	1.0	0.13	1.0	0.13
Ethylene Glycol	25.0	2.69	25.0	2.69
Preservative (57% mercury)	1.8	0.11	1.8	0.11
Rutile Titanium Dioxide				
(nonchalking)	170.0	5.17	170.0	5.17
Calcium Carbonate	253.5	11.23	253.5	11.23

Grind the above materials in a high-speed mill (Cowles, 3800–4500 fpm for 10–15 min) and let down, at a slower speed, as follows:

Rhoplex AC-388 (50%)	459.8	52.23	390.8	44.39
Long Oil Alkyd*	–	–	30.8	3.69
Antifoamer	1.0	0.13	1.0	0.13
Ammonium Hydroxide				
(28%)	2.0	0.27	2.0	0.27
Tributyl Phosphate	11.5	1.41	9.8	1.21
Propylene Glycol	35.0	4.07	35.0	4.07
Water and/or Hydroxy-				
ethyl Cellulose (2.5%)	18.6	2.23	54.8	6.58

* Drier treated with 0.5% of 6% cobalt, 0.5% of 6% manganese, and 1.4% of 24% lead prior to incorporation of alkyd.

	No. 2	No. 3
PVC (%)	40.0	40.0
Total Solids by Vol. (%)	41.0	41.0
pH	≈ 9.4-9.7	≈ 9.4-9.7
Visc.–Initial (KU)	72–75	72–75

No. 4

		lb	gal
Ethylene Glycol		20.0	2.15
Hydroxyethyl Cellulose		1.5	0.13
Water		150.0	18.00
Tamol 850 (30%)		8.6	0.86
KTPP		1.5	0.07
Triton CF-10		2.5	0.28
Dapro 881-S		1.0	0.14
Rutile Titanium Dioxide		150.0	4.56
Zinc Oxide		50.0	1.07
Calcium Carbonate		234.1	10.36

Grind the above materials in a high-speed mill (Cowles) for 10–15 min and let down at a slower speed as follows:

		lb	gal
Rhoplex AC-707 (65%)		345.1	38.48
Texanol	Premix	10.7	1.35
Propylene Glycol	Premix	34.0	3.95
Dapro 881-S	Premix	1.0	0.14
Skane M-8	Premix	2.0	0.23
Ammonium Hydroxide	Premix	2.0	0.27
Water or HEC (2½% sol'n.)		149.4	17.96

Total Solids by Vol. (%)	40.0
PVC (%)	40.0
Visc.—Initial (KU)	80–85
pH	9.6

Acrylic Exterior Sheen Tint Base

	lb	gal
Water	60.0	7.20
Tamol 850 (30%)	15.8	1.58
Triton X-100	2.5	0.28
Experimental Stabilizer 612	3.7	0.48
Propylene Glycol	60.0	6.98
Dapro 881-S	1.0	0.14

Titanium Dioxide	170.0	4.98
Zinc Oxide	25.0	0.54
ASP-170	139.4	6.48

Grind on a high-speed dissolver (Cowles) for 15-20 min at 4000-4400 fpm. Let down at slower speed as follows:

Rhoplex AC-707 (65%)	402.6	44.89
Water	15.0	1.80
Texanol	12.5	1.58
Dapro 881-S	2.0	0.28
Propylene Glycol	38.8	4.51
Skane M-8	2.0	0.23
Natrosol 250 MR (3% sol'n.)	80.0	9.76
Water	69.1	8.29
Visc.–Initial (KU)	72-76	
pH–Initial	9.3-9.6	
PVC (%)	30	
Total Solids by Vol. (%)	40	

Latex Exterior Semigloss Tint Base

Formula No. 1

	lb	gal
Charge the following to the tank:		
Water	2.0	0.24
Propylene Glycol	60.0	6.96
Add the following at low speed, 1900 fpm:		
Tamol 850 (30%)	9.0	0.90
Rutile Titanium Dioxide	225.0	6.57
Barytes	40.0	1.09
Increase speed to 5400 fpm and grind for 20 min; then add the following at low speed, 1900 fpm:		
Propylene Glycol } Premix	120.0	13.92
Preservative } Premix	1.0	0.13
Rhoplex AC-22 (44.5%)	440.0	49.72

Rhoplex AC-61 (46%)		106.0	11.96
Acrysol G-110 (22%)	Premix	10.3	1.23
Water	Premix	46.8	5.66
Triton X-207		10.0	1.22
Triton X-114		1.5	0.17
Colloid 600		2.0	0.26

Total Solids (%)	47.7
PVC (%)	23.1
Visc.–Initial (KU)	75

No. 2

		lb	gal
Charge the following to the tank:			
KTPP	Premix	1.3	0.06
Water	Premix	9.6	1.15
Tamol 850 (30%)		4.0	0.40
Propylene Glycol		67.3	7.82
G.E. Antifoamer 66		0.5	0.06

Add the following at low speed:

Rutile Titanium Dioxide (high gloss)	216.3	6.32
Barytes	38.5	1.05

Grind the above in a high-speed mill (Cowles, 3800–4500 fpm for 10–15 min) and let down at lower speed with the following:

Propylene Glycol		82.2	9.54
Rhoplex AC-490 (46.5%)		509.5	57.58
Acrysol G-110	Premix	11.5	1.36
Water	Premix	84.6	10.15
Super Ad-It	Premix	0.5	0.06
Butyl **Cellosolve**	Premix	23.5	3.00
Triton N-57		9.6	1.12
Triton X-114		2.9	0.33

PVC (%)	23.2
Total Solids by Vol. (%)	31.8
Total Solids by Wt. (%)	48.8
Visc.–Initial (KU)	68

No. 3

		lb	gal
Charge the following to the tank:			
Tamol 731 (25%)		5.0	0.55
Tamol 850 (30%)		4.2	0.42
Propylene Glycol		54.3	6.30
Add the following at low speed, 1900 fpm:			
Rutile Titanium Dioxide		204.0	5.95
Barytes		36.0	0.90
Increase speed to 5400 fpm and grind for 20 min; then add the following at low speed, 1900 fpm:			
Propylene Glycol	Premix	110.0	12.78
Preservative	Premix	1.0	0.13
Rhoplex AC-22 (44.5%)		400.0	45.20
Rhoplex AC-61 (46%)		100.0	11.30
Acrysol G-110 (22%)	Premix	9.3	1.11
Water	Premix	115.0	13.90
Triton N-57		5.4	0.62
Triton X-114		5.4	0.62
Triton X-200 (28%)		0.9	0.10
Colloid 600		3.6	0.47

Total Solids (%)	46
PVC (%)	22.7
Visc.–Initial (KU)	70

No. 4

	lb	gal
Charge the following to the tank:		
Tamol 850 (30%)	9.00	0.91
Water	2.00	0.24
Propylene Glycol	60.00	6.96

Add the following at low speed, 1900 fpm:

Rutile Titanium Dioxide		225.00	6.57
Barytes		40.00	1.09

Increase speed to 5400 fpm and grind for 20 min; then add the following at low speed, 1900 fpm:

Propylene Glycol	Premix	120.00	13.92
Preservative		1.00	0.12
Rhoplex AC-22 (44.5%)		440.00	49.72
Rhoplex AC-61 (46%)		106.00	11.96
Acrysol G-110 (22%)	Premix	10.30	1.22
Water		46.75	5.65
Triton N-57		4.00	0.46
Triton X-207		4.00	0.49
Colloid 600		2.00	0.26

Total Solids (%)	49.03
PVC (%)	22.7
Visc.—Initial (KU)	70

No. 5

		lb	gal
Charge the following to the tank:			
KTPP	Premix	1.3	0.06
Water		10.0	1.20
Tamol 850 (30%)		4.2	0.42
Propylene Glycol		70.0	8.14
Preservative		1.0	0.12

Add the following at low speed, 1900 fpm:

Rutile Titanium Dioxide	225.0	6.57
Barytes	40.0	1.09

Increase speed to 5400 fpm and grind for 20 min; then add the following at low speed, 1900 fpm:

Propylene Glycol	110.0	12.78
Rhoplex AC-22 (44.5%)	440.0	49.72
Rhoplex AC-61 (46%)	110.0	12.40

Nopco NDW		2.0	0.27
Acrysol G-110 (22%)	Premix	12.0	1.42
Water	Premix	68.0	8.21
Triton N-57		10.0	1.16
Triton X-114		3.0	0.34

Total Solids (%)	47
PVC (%)	24
Visc.–Initial (KU)	64

Latex Exterior Light Tint Base

	lb	gal
Pigment grind:		
Water	307.0	37.0
Tamol 731	10.0	1.2
Igepal CTA-639	2.0	0.2
Dowicil S-13	5.0	0.5
Ethylene Glycol	20.0	2.2
Ucar Filmer 351	10.0	1.3
Defoamer*	5.0	0.7
Cellosize Hydroxyethyl Cellulose **QP-4400H**	6.0	0.5
Rutile Titanium Dioxide**	175.0	5.3
Mica (325 mesh, water-ground)	25.0	1.2
Lorite	125.0	5.9
Let down:		
Ucar Latex 365	398.0	44.0

* Examples are **Bubble Breaker 746, Colloid 585.**

** Examples are **Ti Pure R-966, Titanox RA-46.**

PVC (%)	35
Total Solids (%)	51.3
Visc. (KU)	85-90
lb/gal	10.9

Vinyl Ethylene Exterior Tint Base

Natrosol 250 HR (2% sol'n.)		75.0
Tamol 850		6.0
Strodex MOK-70		5.0
Igepal CO-610		4.0
Ethylene Glycol		25.0
Texanol		8.0
Drew L-475		2.0
Dowicil S-13		3.0
Titanox 2060 (CL-NC)		150.0
Zinc Oxide (N.J.Z. #503)		50.0
Minex 4		175.0
Water		95.0
Natrosol 250 HR (2% sol'n.)		150.0
Drew L-475		1.0
Cargill 7700		47.0
Zirco Drier		0.8
Cobalt Drier		1.6
Airflex 720		330.0
Total Solids (%)	53.5	
PVC (%)	37.6	

Acrylic White Exterior Topcoat

Formula No. 1

		lb	gal
Natrosol 250 MHR	Premix	3.0	0.26
Ethylene Glycol	Premix	25.0	2.69
Water		120.0	14.40
Tamol 850 (30%)		10.5	1.05
KTPP		1.5	0.07
Triton CF-10		2.5	0.28
Drew L-475		1.0	0.13
Propylene Glycol		34.0	3.94
Ti Pure R-960		250.0	7.60
AZO-11		50.0	1.07
Asbestine 3X		178.9	7.73

Grind for 10–15 min on Cowles dissolver at 3800-4500 fpm. Let down at a slower speed as follows:

Rhoplex AC-64 (60.5%)	380.0	42.37
Drew L-475	3.0	0.39
Texanol	11.0	1.39
Skane M-8	2.0	0.23
Ammonium Hydroxide	2.0	0.27
Water	116.3	13.95
Natrosol 250 MHR (2½% sol'n.)	20.0	2.44

PVC (%)	40.0
Total Solids by Vol. (%)	41.0
Visc.–Initial (KU)	75–80
pH	9.5

No. 2

(Unmodified)

	lb	gal
Hydroxyethyl Cellulose (2.5% sol'n.)	85.0	10.37
Water	62.5	7.50
Tamol 850 (30%)	10.5	1.05
Triton CF-10	2.5	0.28
KTPP	1.5	0.07
Nopco NXZ	1.0	0.13
Ethylene Glycol	25.0	2.69
Ti Pure R-960	237.5	7.22
AZO-11	50.0	1.07
Asbestine 3X	187.7	8.11

Grind the above materials in a high-speed mill (Cowles, 3800–4500 fpm for 20 min) then add the following at slower speed.

Rhoplex AC-388 (50%)		459.8	52.23
Nopco NXZ		1.0	0.13
Tributyl Phosphate	Premix	11.0	1.36
Propylene Glycol	Premix	34.0	3.95
Skane M-8 (45%)	Premix	2.0	0.23
Ammonium Hydroxide (28%)		1.0	0.13
Water		29.0	3.48

PVC (%)	40.0
Total Solids by Vol. (%)	41.0
Total Solids by Wt. (%)	58.7
pH	9.5

No. 3

(Alkyd Modified)

	lb	gal
Hydroxyethyl Cellulose (2.5% sol'n.)	85.0	10.37
Water	62.5	7.50
Tamol 850 (30%)	10.5	1.05
Triton CF-10	2.5	0.28
KTPP	1.5	0.07
Nopco NXZ	1.0	0.13
Ethylene Glycol	25.0	2.69
Ti Pure R-960	237.5	7.22
AZO-11	50.0	1.07
Asbestine 3X	187.7	8.11

Grind the above materials in a high-speed mill (Cowles, 3800–4500 fpm for 20 min) then add the following at slower speed.

		lb	gal
Rhoplex AC-388 (50%)		390.8	44.39
Long Oil Alkyd (drier treated)		30.8	3.69
Nopco NXZ		1.0	0.13
Tributyl Phosphate	Premix	9.3	1.15
Propylene Glycol	Premix	34.0	3.95
Skane M-8 (45%)	Premix	2.0	0.23
Ammonium Hydroxide (28%)		1.0	0.13
Water		65.3	7.84

PVC (%)	40.0
Total Solids by Vol. (%)	41.0
Total Solids by Wt. (%)	58.5
pH	9.5

No. 4

(Alkyd Modified, High-Hiding)

	lb	gal
Propylene Glycol	48.0	5.58
Hydroxyethyl Cellulose	1.0	0.09
Water	80.0	9.60
Tamol 850 (30%)	11.1	1.11
KTPP	1.5	0.07
Triton CF-10	2.5	0.28
Dapro 881-S	1.0	0.14
Ethylene Glycol	25.0	2.69
Rutile Titanium Dioxide	235.0	7.14
Anatase Titanium Dioxide	15.0	0.46
Zinc Oxide	50.0	1.07
Silica	293.7	13.33

Grind the above materials in a high-speed mill (Cowles) for 10-15 min and let down at a slower speed as follows:

		lb	gal
Rhoplex AC-707 (65%)		403.2	44.96
Long Oil Alkyd (dried treated*)		41.3	4.95
Texanol	Premix	12.3	1.55
Dapro 881-S	Premix	1.0	0.14
Skane M-8	Premix	2.0	0.23
Water or HEC (2½% sol'n.)		55.0	6.61

Total Solids by Vol. (%)	55.0
PVC (%)	40.0
Visc.–Initial (KU)	75-80

* 0.5% of 6% cobalt naphthenate
0.5% of 6% manganese naphthenate
5.6% of 6% zirconium naphthenate

Latex Exterior Chalking White Topcoat

(Alkyd Modified, Zinc Oxide)

	lb	gal
Water	208.00	25.00
Strodex MOK-70	5.00	0.50
KTPP	2.00	0.22
Ethylene Glvcol	32.60	3.50
Texanol	8.00	1.00
Nonmercurial Bactericide	1.00	0.10
Igepal CO-887	4.50	0.50
Nopco NXZ	1.90	0.25
Methocel HG-90	3.00	0.26
Rutile Titanium Dioxide	190.00	5.56
Anatase Titanium Dioxide	30.00	0.93
AZO-33	100.00	2.14
$CaCO_3$	100.00	4.43
Talc	50.00	2.16
Mica (325 mesh, water-ground)	25.00	1.06
Zirconium Drier	1.60	0.16
Cobalt Drier (6%)	1.00	0.10
Antiskin	1.00	0.13
Alkyd 1271 (100%)	48.00	5.75
Igepal 887	2.00	0.25
Latex (55% solids)	326.00	36.00
Water	82.30	9.88
Nopco NXZ	1.00	0.13

lb/gal	12.24
PVC (%)	39.4
Visc.–Initial (KU)	78

Vinyl Acrylic Exterior White Topcoat

(With Zinc Oxide)

Water	168.0
Daxad 30	11.0
Triton CF-10	2.0

KTPP	2.0
Nopco NDW	1.0
Cosan 171-S	10.0
Titanox RA-NC	250.0
Zinc Oxide	25.0
Asbestine 3X	175.0

Mix on high-speed mill and add:

Texanol	7.0
Ethylene Glycol	25.0
Natrosol 250 MR	2.0
T&W 345-100 Super Alkyd	30.0
Lead (24%)	1.0
Manganese (6%)	0.5
Nopco NDW	1.0
ParCryl 400	395.0
Ammonium Hydroxide	2.0
Water	44.0

PVC (%)	30
Total Solids (%)	55
Visc. (KU)	74–80
pH	9.5

Latex Exterior Pigmented Topcoat

Multron Mill Base:

Multron R-12A	44
Multron R-221	177
EAB 381-1/10 (10% in ethylglycol acetate)	18
Vinylite VAGH (15% in ethylglycol acetate)	212
Modaflow (10% in ethyl acetate)	2
Titanium Dioxide **R-KB-2**	192
Bentone 27 (10% in ethylglycol acetate)	44
Ethyl Glycol Acetate	79
Xylol	20
Solvesso 100	20
Pentoxone	20

Mondur Sol'n.:

Mondur CB-75		172
Total Solids (%)	58	
P/B Ratio	0.5/1	

To obtain spray viscosity, dilute with approximately 500 parts of the following solvent blend—ethyl acetate, butyl acetate, ethyl glycol acetate, toluene (1:1:1:1 pbw). After about 6-months exterior exposure, chalking will be noted.

To obtain better chalking and yellowing resistance, 195 parts of **Mondur HC** may be substituted for the 172 parts **Mondur CB-75**.

Acrylic Exterior Trim Paint

Formula No. 1

(Gloss)

	lb	gal
Propylene Glycol	60.0	6.98
Tamol 731 (25%)	14.9	1.62
Nopco NXZ	1.0	0.13
Ti Pure R-900	275.0	8.05

Grind the above materials in a high-speed mill (Cowles, 3800–4500 fpm for 20–25 min) then add the following at slower speed:

	lb	gal
Water	66.5	7.98
Rhoplex AC-507	546.7	62.38
Nopco NXZ	1.0	0.13
Texanol	25.0	3.15
Skane M-8	2.0*	0.23
Propylene Glycol	40.0	4.64
Ammonium Hydroxide (28%)	2.0	0.27
Formaldehyde (37%)	4.0**	0.44
Hydroxyethyl Cellulose (3% sol'n.)	33.5	4.00

PVC (%)	23.0
Total Solids by Vol. (%)	35.0

Total Solids by Wt. (%)	49.5
pH	8.6-9.1
Visc. (KU)	75-81

---

* Level can be increased to 4 lb/100 gal without change in levels of formaldehyde or ammonia.

** If pH of paint exceeds 9.1 additional 1.0 lb/100 gal of formaldehyde should be added.

No. 2

(Semigloss)

		lb	gal
Water		43.1	5.17
Propylene Glycol		40.0	4.65
Drew 475S		1.0	0.13
Tamol 731 (25%)		10.0	1.09
Triton CF-10		2.0	0.22
DMAE		2.0	0.24

Mix, then add the following:

		lb	gal
Zopaque RCL-9		250.0	7.30
ASP-170		58.0	2.70

Grind at high speed, then let down at low speed:

		lb	gal
Rhoplex AC-64 (60.5%)		463.8	51.00
Propylene Glycol	Premix	35.0	4.07
Texanol	Premix	14.0	1.77
Formaldehyde	Premix; raise pH to 8 with ammonia	3.0	0.33
Skane M-8	Premix; raise pH to 8 with ammonia	2.0	0.23
Drew 475S		3.0	0.41
Natrosol 250 MR (2½%)	Adjust ratio to give indicated viscosity	125.0	15.00
Water	Adjust ratio to give indicated viscosity	47.4	5.69

PVC (%)	25
Total Solids by Vol. (%)	40
Visc.–Initial (KU)	72–75
pH	8.7-9.0

No. 3

(Semigloss, White)

Water		43.1
Propylene Glycol		40.0
Colloid 643		2.0
Tamol 731 (25%)		10.0
Triton CF-10		2.0
DMAE		2.0

Mix, then add the following:

Zopaque RCL-9	250.0
ASP-170	58.0

Grind at high speed, then let down at low speed:

Rhoplex AC-64 (60.5%)		463.8
Colloid 643		6.0
Formaldehyde	Premix	3.0
Water	Premix	30.0
Propylene Glycol	Premix	35.0
Texanol	Premix	14.0
Skane M-8	Premix	2.0
Natrosol 250 MR (2½%)	Adjust ratio to give indicated viscosity	142.4
Water	Adjust ratio to give indicated viscosity	

PVC (%)	25
Total Solids by Vol. (%)	40
Visc.—Initial (KU)	72–75
pH	8.7-9.0

No. 4

(White, Gloss)

	lb	gal
Propylene Glycol	75.0	8.72
Tamol 731 (25%)	15.0	1.63
Triton CF-10	1.0	0.11
Antifoamer	1.0	0.13

Water		30.0	3.60
Rutile Titanium Dioxide (nonchalking, high gloss grade)		275.0	8.06

Grind the above materials in a high-speed mill (Cowles, 3800–4500 fpm for 20–25 min) and let down, at a slower speed, as follows:

Rhoplex AC-388 (50%)		503.6	57.21
Tributyl Phosphate or **Texanol**	Premix	3.0	0.37
Antifoamer	Premix	1.0	0.13
Super Ad-It	Premix	9.0	1.11
Ammonium Hydroxide (28%)	Premix	2.0	0.27
Propylene Glycol	Premix	45.0	5.23
Water	Premix	102.5	12.30
Acrysol G-110 (22%)	Premix	10.0	1.13

PVC (%)	23.0
Total Solids by Vol. (%)	35.0
Total Solids by Wt. (%)	49.1
pH	≈ 9.4–9.7
Visc.—Initial (KU)	≈ 65–70

No. 5

(White, Gloss)

		lb	gal
Propylene Glycol	Premix	75.0	8.72
Hydroxyethyl Cellulose (100%)	Premix	1.2	0.10
Tamol 731 (25%)		17.2	1.87
Triton CF-10		1.0	0.11
Antifoamer		1.0	0.13
Water		40.0	4.80
Rutile Titanium Dioxide (nonchalking, high gloss)		250.0	7.55

Grind the above materials in a high-speed mill (Cowles, 3800–4500 fpm for 20-25 min) and let down, at a slower speed, as follows:

Rhoplex AC-388 (50%)	395.2	44.89
Rhoplex AC-61 (46%)	185.9	20.97

Tributyl Phosphate	Premix	7.2	0.89
Antifoamer	Premix	2.0	0.26
Super Ad-It	Premix	9.0	1.11
Ammonium Hydroxide (28%)	Premix	2.0	0.27
Water		69.4	8.33

PVC (%)	20.0
Total Solids by Vol. (%)	37.75
Total Solids by Wt. (%)	50.5
pH	≈ 9.4-9.7
Visc.–Initial (KU)	≈ 65-70

No. 6

(White)

		lb	gal
Charge the following to the tank:			
Propylene Glycol		60.0	6.98
Tamol 850 (30%)		9.2	1.01
Nopco NXZ		1.0	0.13
Water		5.1	0.62
Triton X-100		2.5	0.28
Experimental Stabilizer 612		3.7	0.48
Rutile Titanium Dioxide (high-gloss grade)		275.0	8.05
Zinc Oxide		10.0	0.21

Grind the above materials in a high-speed mill (Cowles, 3800-4500 fpm for 20-25 min) and let down at a slower speed as follows:

Water		38.0	4.56
Rhoplex AC-507 (46.5%)		546.7	62.38
Texanol	Premix	25.0	3.16
Nopco NXZ	Premix	1.0	0.13
Propylene Glycol	Premix	36.0	4.19
Skane M-8	Premix	4.0	0.46
Natrosol 250 MR (3%)/Water		65.7	7.36

PVC (%)	23.4
Total Solids by Vol. (%)	35.2
Total Solids by Wt. (%)	50.0
pH	9.8
Visc. (KU)	72-76

No. 7

(Tintable White)

Charge the following to the tank:

Propylene Glycol	70.0
Tamol 731 (25%)	14.9
Nopco NXZ	1.0
Water	20.0
Rutile Titanium Dioxide (high-gloss grade)	275.0

Grind the above materials in a high-speed mill (Cowles, 3800-4500 fpm for 20--25 min) and let down at a slower speed as follows:

Water		47.7
Rhoplex AC-507 (46.5%)		546.7
Texanol or Tributyl Phosphate	Premix	18.4
Super Ad-It	Premix	9.0
Nopco NXZ	Premix	1.0
Ammonium Hydroxide (28%)		2.0
Propylene Glycol		30.0
Acrysol G-110 (22%)	Premix	10.0
Water	Premix	23.5
Oleic Acid		2.0

PVC (%)	23.0
Total Solids by Vol. (%)	35.0
Total Solids by Wt. (%)	49.5
pH	9.4-9.7
Visc. (KU)	76-82

No. 8

(Green Deep Tone)

Charge the following to the tank:

Propylene Glycol	68.5
Tamol 731 (25%)	3.5
Nopco NXZ	1.0
Water	21.5
Triton X-207	1.7
Hydrite UF	70.1

Grind the above materials in a high-speed mill (Cowles, 3800–4500 fpm for 20–25 min) and let down at a slower speed as follows:

Water		68.5
Rhoplex AC-507 (46.5%)		544.5
Texanol	Premix	13.0
Super Ad-It		8.9
Nopco NXZ		1.0
Ammonium Hydroxide (28%)		2.0
Aurasperse W-6017 (chrome oxide green)		85.4
Acrysol G-110 (22%)*	Premix	64.9
Water		

---

* To adjust initial viscosity to 75–80 KU.

PVC (%)	14.4
Total Solids by Vol. (%)	31.4
Total Solids by Wt. (%)	38.4
pH	9.3–9.7
Visc. (KU)	75–80

No. 9

(Blue Deep Tone)

Charge the following to the tank:

Propylene Glycol	69.0
Tamol 731 (25%)	4.0
Nopco NXZ	1.0

Water	20.0
Triton X-207	2.0
Hydrite UF	38.1
Rutile Titanium Dioxide	20.0

Grind the above materials in a high-speed mill (Cowles, 3800–4500 fpm for 20–25 min) and let down at a slower speed as follows:

Water		35.1
Rhoplex AC-507 (46.5%)		548.0
Texanol	Premix	13.0
Super Ad-It	Premix	9.0
Nopco NXZ	Premix	1.0
Ammonium Hydroxide (28%)		2.0
Aurasperse W-4123 (phthalo blue)		80.3
Natrosol 250 MR (3%)* / Water	Premix	65.6

---

* To adjust initial viscosity to 75–80 KU.

PVC (%)	11.6
Total Solids by Vol. (%)	30.5
Total Solids by Wt. (%)	36.1
pH	9.3-9.7
Visc. (KU)	75-80

No. 10

(Yellow Deep Tone)

Charge the following to the tank:

Propylene Glycol	68.6
Tamol 731 (25%)	3.0
Nopco NXZ	1.0
Triton X-207	1.7
Aurasperse W-1021 (yellow iron oxide)	139.3
Hydrite UF	52.3

Grind the above materials in a high-speed mill (Cowles, 3800–4500 fpm for 20–25 min) and let down at a slower speed as follows:

Water		53.5
Rhoplex AC-507 (46.5%)		545.4
Texanol	Premix	12.8
Super Ad-It	Premix	9.0
Nopco NXZ	Premix	2.0
Ammonium Hydroxide (28%)		2.0
Acrysol G-110 (22%)*	Premix	60.2
Water	Premix	

* To adjust initial viscosity to 75–80 KU.

PVC (%)	14.5
Total Solids by Vol. (%)	31.4
Total Solids by Wt. (%)	38.5
pH	9.3-9.7
Visc. (KU)	75-80

No. 11

(Red Deep Tone)

Charge the following to the tank:

Propylene Glycol	69.4
Tamol 731 (25%)	3.5
Nopco NXZ	1.0
Water	21.8
Triton X-207	1.7
Hydrite UF	82.7

Grind the above materials in a high-speed mill (Cowles, 3800–4500 fpm for 20–25 min) and let down at a slower speed as follows:

Water		66.7
Rhoplex AC-507 (46.5%)		550.7
Texanol	Premix	13.1
Super Ad-It	Premix	9.1
Nopco NXZ	Premix	1.0
Ammonium Hydroxide (28%)		2.0

Aurasperse W-3040 (red iron oxide)		62.3
Acrysol G-110 (22%)* / Water	Premix	60.8

* To adjust initial viscosity to 75–80 KU.

PVC (%)	14.8
Total Solids by Vol. (%)	31.9
Total Solids by Wt. (%)	39.9
pH	9.3-9.7
Visc. (KU)	75-80

No. 12

(Black Deep Tone)

Charge the following to the tank:

Propylene Glycol	68.8
Tamol 731 (25%)	3.5
Nopco NXZ	1.0
Water	19.9
Triton X-207	1.7
Aurasperse W-7012 (carbon black)	17.2
Hydrite UF	80.8

Grind the above materials in a high-speed mill (Cowles, 3800–4500 fpm for 20–25 min) and let down at a slower speed as follows:

Water		86.4
Rhoplex AC-507 (46.5%)		546.0
Texanol	Premix	12.9
Super Ad-It	Premix	8.9
Nopco NXZ	Premix	1.0
Ammonium Hydroxide (28%)		2.0
Acrysol G-110 (22%)* / Water	Premix	57.0

* To adjust initial viscosity to 75–80 KU.

PVC (%)	12.9
Total Solids by Vol. (%)	30.8
Total Solids by Wt. (%)	37.4
pH	9.3-9.7
Visc. (KU)	75-80

Latex Exterior Trim Paint

Formula No. 1

(Semigloss Green)

	lb	gal
Strodex PK-90	3.3	0.3
Tween 20	2.5	0.2
Ethylene Glycol	106.0	11.4
Foamkill 639	1.6	0.2
Chromium Oxide Green 7099	91.8	2.1
Duramite	101.2	4.5
Barytes	83.5	2.2

Blend the above with slow agitation into a slurry and make one pass through a Morehouse or sand mill to achieve desired grind. Then let down with following under slow agitation (200-300 rpm).

Cyanamer P-250 (2½% sol'n.)	125.2	15.0
Butyl Oxitol	16.7	2.1
Thickener L	21.7	2.4
Water	16.7	2.0
Cellosize WP-4400 (2½% sol'n.)	5.0	0.6
Geon 450x20	242.8	25.8
Arolon 363	293.8	33.9
Colloid 600	0.8	0.8

lb/gal	10.74
PVC (%)	23.28
Total NV by Wt. (%)	50.62
Total NV by Vol. (%)	36.52
pH	7.3
Visc. (KU)	75

No. 2

(Semigloss)

	lb	gal
Premix:		
Water	108.3	13.00
Daxad 30	8.0	0.94
Tergitol NPX	5.0	0.56
Victawet 35-B	3.0	0.31
Increase speed and add slowly:		
Chemacoil TA-100	78.0	10.00
Duramac 2438	116.0	14.53
Adjust speed to disperse pigments:		
Ti Pure R-901	300.0	8.76
Cellosize QP-15000 (dry)	0.5	—
Polyglycol P-1200	6.0	0.71
Ethylene Glycol	37.0	4.00
Nuodex PMA-18	4.5	0.41
Colloid 581-B	1.0	0.15
Slow to mixing speed:		
Everflex BG (52%)	437.0	49.10
Super-Cobalt Drier	1.0	0.13

Visc.—Initial (KU) 72

Alkyd Red Shingle Stain

(Thixotropic)

	lb	gal
Red Iron Oxide	275.00	6.88
Aroplaz 1254-M-70	177.00	22.21
Thixcin R	20.00	2.46
Mineral Spirits	150.00	23.00

Mineral Spirits	301.00	46.20
Lead Naphthenate (24%)	2.72	0.28
Cobalt Naphthenate (6%)	1.00	0.13

PVC (%)	32.7
Total NV by Vol. (%)	19.6
lb/gal	9.17

Procedure:

Mix iron oxide, **Thixcin R**, and **Aroplaz 1254** thoroughly. Add mineral spirits slowly to a smooth paste free from lumps. If properly mixed one roll mill pass will give a smooth paste. Care should be taken to add small amount of mineral spirits in the thin-down to avoid lumping.

This formula has excellent application properties especially for shake type shingle.

Note:

Thixcin R should not be milled at a temperature higher than 130 F to avoid seeding.

Whitewash

Formula No. 1

Casein	2.6
Hydrated Lime	29.0
Trisodium Phosphate	2.6
Water	65.8

No. 2

Common Salt	12 lb
Powdered Alum	6 oz
Molasses	1 qt
Lime Paste	8 gal

Rust-Inhibitive Wash

Orthophosphoric Acid	7.00
Sodium Dichromate	0.75
Caramel Sol'n.	0.50
Cellosolve	0.50
Water	91.25

Vinyl Acetate Exterior Textured Finish

(Roller Applied)

Mix together:

Vinamul 3401	250.0
Courlose F350 (5% aq. sol'n.)	33.0
Potassium Hydroxide (50% aq. sol'n.)	1.0
Butyl Dioxitol Acetate	15.0
Acticide MPM	0.4

Add and disperse:

Tioxide RCR2	50.0

Add and stir thoroughly:

Sodium Benzoate	4.0
Fordacal M2	180.0
Fordacal 400	170.0

Followed by:

Fordacal Z3	174.6
Fordacal Z5	120.0

Finally add under stirring:

Foamaster NDW	2.0

P/B Ratio 5.6:1

This system is of the type widely used in Continental Europe for decorating fair-faced concrete; it is designed for application by patterned roller, and gives an average peak height up to 1.5 mm.

Chapter III

METAL PAINTS

Aside from the specific substrate preparations indicated for each metal or alloy, the most important distinction among metal paints is their drying and curing schedules. Two main groups are air-dry solvent or water systems and bake preparations. The latter group is formulated for and serves industrial needs.

Acrylic-Nitrocellulose Aircraft

(White, Gloss)

RBH White Titanium Dioxide Dispersion No. 6610	23.80
Dioctyl Phthalate	5.40
½″ RS Nitrocellulose (70% in ethyl alcohol)	5.90
Acryloid B-82 (50% solids in toluene)	28.64
Toluene	7.16
Methyl Ethyl Ketone	13.10
Methyl Isobutyl Ketone	16.00

Acrylic Camouflage Lacquer

(Gull Gray)

Ti Pure R-610	12.55
Celite 266	1.13
Talc #399	3.17
Acryloid B-82 (50% solids in toluene)	25.60
Dioctyl Phthalate	4.65
½″ RS Nitrocellulose (70% in ethyl alcohol)	8.31

Toluene	9.19
Methyl Isobutyl Ketone	17.70
Methyl Ethyl Ketone	17.70

Procedure:

Grind for 24 h in a pebble mill and tint to proper color with RBH Black Dispersion No. 6077.

Acrylic Gray Automotive Lacquer

	lb	gal
Roller mill grind:		
Rutile Titanium Dioxide	104.14	2.98
Lampblack	1.09	0.07
Acryloid A-21LV (30%)	105.53	13.25
Cellosolve Acetate	23.40	2.88
Mix with:		
Acryloid A-21LV (30%)	488.53	58.85
Santicizer 160	73.77	7.97
Cellosolve Acetate	30.48	3.77
Methyl Ethyl Ketone	35.57	5.32
Toluene	35.57	4.91

The formulation above is at 40% solids. It should be reduced for spraying to 22% solids with the following solvent mixture:

Acetone	280.0	42.5
Toluene	307.7	42.5
Cellosolve Acetate	121.5	15.0

	(Base Lacquer)	*(Lacquer at Spray Weight)*
lb/gal	8.78	7.93
Total Solids (%)	40	22.1
P/B Ratio	30:70	30:70

Acrylic Exterior Can Enamel

Formula No. 1

(Clear)

	lb	gal
Ubatol A-745	390.00	44.70
Water	314.00	37.70
Dimethylamino Ethanol	9.40	1.17
Butyl **Cellosolve**	47.00	6.21
Texanol	39.00	4.94
Cymel 1123	17.20	1.72
Resimene X-735	24.60	2.38
#3 Carnauba Wax 15625-S	6.50	0.77
ERL-4221	3.98	0.41

pH	9.0
Visc. (KU)	70
Total Solids by Wt. (%)	23
Total Solids by Vol. (%)	20
lb/gal	8.52

No. 2

(Clear)

	lb	gal
Ubatol A-760S	679.00	80.00
Carnauba Wax (25%)	7.80	0.92
Dimethylamino Propanol	8.20	1.09
Butyl **Cellosolve**	37.20	4.90
Texanol	5.70	1.98
Water	46.90	5.63
Cymel 1123	57.80	5.78

Total Solids by Wt. (%)	27.0
Total Solids by Vol. (%)	25.5
lb/gal	8.42
pH	8.7
Visc. (s–Ford No. 4 Cup)	40–45

No. 3

(White)

	lb	gal
Ubatol A-760-S	176.00	20.40
Dimethylamino Ethanol	2.61	0.30
Emultex R	3.30	0.38
Hexylene Glycol	47.30	6.15

Mix well, then add:

	lb	gal
R-902	244.00	7.00

High speed grind 10 min.
Let down, premix before adding grind.

	lb	gal
Ubatol A-760S	322.00	37.40
Dimethylamino Ethanol	3.77	0.41
Cymel 1123	49.60	4.96
Carnauba Wax (25%)	5.95	0.70
Texanol	23.70	3.04
Water	160.80	19.26

Total Solids by Wt. (%)	40.5
Total Solids by Vol. (%)	24.0
lb/gal	10.4
pH	8.7
Visc. (s–Ford No. 4 Cup)	40-50

Acrylic Interior Can Coating

P2-103 Acrylic Resin	321.88
Cymel 303 Methylated Melamine Resin	57.69
Ammonia (28% sol'n.)	18.03
Butyl **Cellosolve**	30.00
Ethyl Alcohol (35%)	20.00
Water	547.59
Carnauba Wax Dispersion (20%) (See below for preparation)	4.81

Thoroughly mix all components. Reduce 4:1 with water. Make application with #14 wire wound bar. Baking cycle is 10 min at 400 F. Dry film weight should adjust to 4.5 mg/in.^{2} There is no flash time.

To prepare Carnauba wax dispersion (20%):

Charge into mixing tank:

P2-103 Acrylic Resin	197

Premix the following and add to the above while mixing. Mix well, then charge to pebble mill.

Demineralized Water	103
Aqua Ammonia (26° Bé)	11
Ethanol (methanol denatured)	40
Ethylene Glycol Monobutyl Ether	40

Add water and wax to the above in pebble mill. Grind for 72 h.

Water	409
#2 Yellow Carnauba Wax (powdered)	200

Check for cratering in coating to be used.

Epoxy Interior-Exterior Can Coating

(Unpigmented)

	lb	gal
Epon 1001-B-80	246.01	27.19
Toluene	29.25	4.04
Methyl Butyl Ketone	21.52	3.19
n-Butanol	36.90	5.45
Cellosolve	78.51	10.13
Uni-Rez 2100P75	249.61	31.53
n-Butanol	47.05	6.95
n-Propanol	62.04	9.26
Toluene	16.36	2.26

Application:

This coating offers excellent long term protection to internal and external surfaces of aluminum, tinplate, and steel cans. Recommended application is by spray, dipcoat, or roller.

Phenolic Interior-Exterior Can Coating

Bakelite 10282 (pure phenolic)	100.0
Admolene (catalyzed linseed)	97.0
Xylene	223.0
Zinc Naphthenate	2.0
Manganese Naphthenate	0.3
Cobalt Naphthenate	0.8

Alkyd Coil Coating Enamel

(White)

Titanium Dioxide	300
M-P-A 60 (xylene)	6
Beckosol 93-642	489
Super Beckamine 27-580	105
Solvesso 150	236
Butyl **Cellosolve**	39
Butyl **Carbitol**	50

Visc.–Initial (s–No. 4 Ford Cup)	50–60
lb/gal	12.3

Polyester White Coil Coating

Formula	No. 1 *(High Gloss)*	No. 2
	lb/100 gal	
Unitane OR-650	248.1	247.0
Cyplex 1531 (60%)	137.8	–
Cyplex 1538 (65%)	–	126.7
Polyethylene 6A Paste (30% in xylol)	19.6	19.2

Disperse on 3-roll mill and add:

Cyplex 1531 (60%)	449.2	–
Cyplex 1538 (65%)	–	412.9
Cymel 303 (100%)	39.1	39.1

Cellulose Acetate Butyrate (25% sol'n.)	62.6	62.6
Catalyst 1010	2.9	2.1
Ethylene Glycol Monoethyl Ether Acetate (EGMEA)	3.8	3.5
Isophorone	7.8	7.6
Solvesso 150	54.3	92.1
Butanol	24.8	24.6
Total NV by Wt. (%)	63.0	63.6
Total NV by Vol. (%)	47.3	48.7
P/B Ratio by Wt.	60:100	60:100
PVC (%)	15.7	15.2
Visc. (s–Ford Cup No. 4)	107	100

Application: Reverse roll coater

No. 3

(High Gloss)

	lb/100 gal
Unitane OR-650	331.6
Cyplex 1546 (70%)	188.8

Disperse on 3-roll mill and add:

Cyplex 1546	332.0
Cymel 303 (100%)	39.8
Cymel 247-10 (60%)	20.9
Catalyst 1010	6.3
Silicone **L5310** Resin	0.4
Ethylene Glycol Monoether Acetate	52.4
Diacetone Alcohol	31.4
Isophorone	7.1
Ethylene Glycol Monobutyl Ether	8.3
Solvesso 150	56.2
n-Butanol	36.1

Total NV by Wt. (%)	67.5
Total NV by Vol. (%)	51.6
P/B Ratio by Wt.	80:100
PVC (%)	19.3

	No. 4 *(Flat)*	No. 5 *(Semigloss)*	No. 6 *(High Gloss)*
	←	lb/100 gal	→
Unitane OR-640-A	223.9	261.8	283.5
Asbestine 3X	44.8	22.4	–
Fibrene C400	179.1	89.8	–
Cyplex 1526 (50%)	134.0	127.6	188.9
Solvesso 150	140.8	73.5	–
Polyethylene 6A (30% in xylol)*	3.6	13.6	15.5
Disperse and add:			
Cyplex 1526 (50%)	265.3	361.2	369.7
Cymel Hexamethoxymethylmelamine	22.1	27.1	31.0
Catalyst 1010	3.4	4.1	4.9
Isophorone	22.4	21.7	21.7
Solvesso 150 Solvent	37.5	41.2	78.9
Butanol	42.0	40.1	39.7
Total Solids by Wt. (%)	60.0	60.0	58.0
Total Solids by Vol. (%)	39.2	41.1	40.8
PVC (%)	41.1	30.3	20.3

Note:

Disperse flat and semigloss formulations with pebble mill; high gloss with roller mill.

Anticratering agents are not necessary under normal application conditions over clean substrates.

---

* Polyethylene 6A should be dissolved in xylol at 30% solids at about 200 F to form a paste when cool. It provides mar-proofing properties.

	No. 7	No. 8
	(Glossy White)	
Composition of enamel (% by wt.):		
Nonvolatile	65.0	58.0
Volatile	35.0	42.0

Composition of nonvolatile (%):		
P/B Ratio	47.3:52.7	47.3:52.7
Composition of pigment (%):		
Rutile Titanium Dioxide	100.0	100.0
Composition of binder (%):		
Cyplex 1476-8 (solids)	85.0	–
Cyplex 1526 (solids)	–	90.0
Cymel Hexamethoxymethylmelamine	15.0	10.0
Composition of volatile (%):		
Solvesso 150 Solvent	70.0	85.0
Ethyl **Cellosolve** Solvent	10.0	–
Butanol	10.0	10.0
Isophorone	10.0	5.0
Additives (% on total resin solids):		
Catalyst 1010	1.0	1.5
Polyethylene No. 6A	–	1.5

No. 9

(White, With Urea-Formaldehyde)

	lb	gal
Unitane OR-600	293.1	8.57
Cyplex 1473-5 (65%)	150.2	16.87
Xylene	18.0	2.49
Disperse above pigment paste:		
Cyplex 1473-5 (65%)	199.9	22.46
Beetle 80	97.7	10.79
2-Ethoxyethylacetate	20.6	2.54
Xylene	138.8	19.23
n-Butanol	112.8	16.70
Paint Additive **DC #1**	1.0	0.13
Catalyst 4040	1.6	0.20

Acrylic Aerosol Metal Paint

Formula	No. 1 *(Clear, Fluorocarbon)*	No. 2 *(Clear, Hydrocarbon)*
Lacquer Mix:		
Acryloid B-66 (50%)	40.0	35.00
Toluene	52.0	8.75
Xylene	–	42.25
Cellosolve Acetate	8.0	14.00
Spray Can (6 oz) Charge		
Lacquer	80.0 g	70.0 g
Freon 11	60.0 g	–
Freon 12	60.0 g	–
Methylene Chloride	–	21.0 g
Isobutane	–	42.0 g
Propane	–	7.0 g
Total Solids (%)	8.0	7.0

No. 3

(White, Fluorocarbon Propellant)

Roller mill grind:	
Ti Pure R-902	7.0
Acryloid B-48N (45%)	7.0
Cellosolve Acetate	1.2
Let down:	
Acryloid B-48N (45%)	22.0
Acetone	25.1
Toluene	27.0
Cellosolve Acetate	10.6
Raybo 3	0.1

Spray can (6 oz) charge:

Above lacquer		80.0 g
Freon 11		60.0 g
Freon 12		60.0 g
Total Solids (%)	8.0	

Alkyd Airless Spray Metal Paint

(Gray)

	lb	gal
Ti Pure R-610	143.3	4.10
Furnace Black	2.8	0.19
Phthalocyanine Blue	0.3	–
Arolon 376-WA8-50	55.5	6.37
Water	111.0	13.33
Odorless Mineral Spirits	5.5	0.87
Cobalt Naphthenate (6%)	1.5	0.18
Lead Naphthenate (24%)	0.4	0.04

Pebble mill – 18 h

	lb	gal
Arolon 376-WA8-50	123.3	14.18
Water	92.5	11.13

Mix and add

	lb	gal
Arolon 376-WA8-50	191.1	21.97

Mix and add

	lb	gal
Octyl Alcohol	9.9	1.44
Water	218.2	26.20

Total Solids (%) 34.8

Alkyd Aluminum Paint

Formula No. 1

(Heat Resistant, Aerosol)

Material Concentrate:

Aroplaz 7323 (60% solids)	55.0
SR-82 (60% solids)	18.0
Catalyst (see Procedure below)	2.0
Aluminum Paste **#1571**	25.0

Procedure:

Catalyst prepared as follows:

Cobalt Octoate (6%)	21.7
Manganese Naphthenate (6%)	21.7
Xylene	23.9
n-Butanol	32.7

Blend all ingredients of the Concentrate together. Be sure to use sufficient agitation to thoroughly disperse the pigment to avoid aggregates which might cause valve plugging.

Aerosol Loading:

Concentrate	15.0
Toluene	40.0
Propellent 12	45.0

Valve and actuator selection:

The B-14-70 valve and 110-20-18 sprayhead manufactured by Newman Green, 151 Interstate Road, Addison, Illinois, has proven satisfactory for this formulation.

Application:

The can should be held 8–12 in. from the object being sprayed. Use oscillating motion while spraying to insure a uniform, smooth coating. Overspraying in one place causes sags and runs. This formula, because it utilizes a leafing pigment, has a silver colored coating but will show some slight pigment rub-off unless cured by heat. Curing during use is sufficient and at temperatures from 300–400 F, this occurs quite rapidly.

No. 2

(Heat-Resistant)

	lb	gal
Alcoa 422* (aluminum powder)	245.00	11.70
Varkyd 385-50X Silicone Alkyd	300.00	38.90
Mineral Spirits	58.00	9.00
Xylol	298.00	41.59
Cobalt Naphthenate (6%)	1.95	0.24
Zirco Drier (6%)	7.25	1.02
Antiskinning Agent	0.44	0.05

lb/gal	8.8
PVC (%)	39
Visc. (KU)	55
Total NV (%)	43

---

* The use of a paste like **Alcoa 1570** yields a paint of somewhat higher consistency. The use of the powder has been recommended as a means of minimizing salt spray test inconsistencies.

Alkyd Metal Paint

	lb/100 gal
Micaceous Iron Oxide	540.00
Zinc Oxide	52.00
724 Black Iron Oxide	6.50
Aluminum Paste (65% NV)	26.00
#38 **Bentone** Mastergel (15%)	78.00
B-600-60 Soya Alkyd	160.00
Mineral Spirits	107.00

Disperse in a Hockmeyer. Add:

B-600-60 Soya Alkyd	198.00
Mineral Spirits	69.00
Troykyd–(6% cobalt naphthenate)	2.40

Troykyd–(5% calcium naphthenate)	0.78
Troykyd–(24% lead naphthenate)	5.85
Troykyd–Antiskinning Agent	2.60
Total Solids by Vol. (%)	45.2
PVC (%)	30
lb/gal	12.48
Visc. (KU)	70–75

4 mil (dry) film coat over 1 mil (dry) anticorrosive primer suggested.

Alkyd Weldable Metallically Pigmented Coating

Phenolic Modified Alkyd (50% in xylene)	19.40
M-P-A (mineral spirits)	1.95
Xylene	18.84
Aluminum Paste (74% in mineral spirits)	2.65
Cobalt Naphthenate	0.38
Manganese Naphthenate	0.40
Lithium Oxide	0.10
Silica Gel	0.47
Zinc Dust (avg. size 6 microns)	55.81

Chlorinated Rubber Steel Paint

	Formula No. 1 *(Primer)*	No. 2 *(Midcoat)*	No. 3 *(Top-coat)*
Alloprene X-10	16.4	17.3	20.8
Cereclor 42P	8.2	8.6	10.0
Cereclor 70S	–	4.0	5.0
Paraplex G-62	0.4	0.5	0.4
Propylene Oxide	0.2	0.2	0.2
Thixatrol ST	0.8	1.1	–
Benetone 34	–	–	0.5
Zinc Chromate	15.7	–	–
Titanium Dioxide	4.2	10.4	20.0

Carbon Black	–	0.1	–
International 3X	–	12.5	–
Mica 325	–	2.0	–
Barytes	15.6	–	–
Xylene	38.5	34.6	28.8
Aromatic 100	–	8.7	14.3
lb/gal	11.5	10.4	10.4
Total NV by Wt. (%)	59.8	56.5	56.7
Total NV by Vol. (%)	38.5	37.2	37.6
PVC (%)	32.6	26.5	16.3
Visc. (KU)	82	–	78

Metal Paint

(Semigloss, Heat-Resistant)

Material Concentrate:

A	**SR-111** (50% silicone)	35.3
	SR-82 (60% silicone)	5.0
	Ethyl Cellulose (7% in xylene)	45.0
	Shawinigan Black Pigment	2.2
	Cab-O-Sil M-5	0.5
	n-Butanol	2.0
	Xylene	9.6
B	**Catalyst SC-3900**	0.4

Procedure:

Blend all ingredients of Part A together. Ball mill until particle size is 6 or less on the Hegmen Grind Scale. Add Part B catalyst.

Aerosol Loading:

Concentrate	20.0
Toluene	25.0
1,1,1 Trichloroethane	10.0
Propellent 12	45.0

Valve, Actuator, and Can Selection:

The B-14-70 valve and B-110-20-18 sprayhead manufactured by Newman Green, 151 Interstate Road, Addison, Illinois, has proven satisfactory for this formulation. Lined cans should be used to insure no effect of the solder on the resin stability.

Application:

Shake can well to disperse pigment. The can should be held 10-12 in. from the object being sprayed. Use oscillating motion while spraying to insure a uniform, smooth coating. The coating dries tack free quite rapidly. Curing during use will usually suffice although the best results are obtained by precuring 30 min at 350–400 F.

Chapter IV

MASONRY PAINTS

A good masonry paint is specifically designed to handle the alkaline conditions in uncured cementitious substrates. The surface requires a high-binder primer because of its porous nature and resultant tendency to suck in liquids.

Weather Barrier Mastic Coating

(Latex)

This weather barrier mastic is especially useful as a sealant and decorative coating for cinder block, brick, concrete, and also for insulation substrates, such as asbestos, silicate, glass, rigid and flexible foams, and metal including aluminum, stainless and galvanized steel. The formulation can be applied by brush, trowel, spray, or palming. Usually films of 1/32- or 1/16-in. thickness are applied over the insulator, but thicker layers can be applied without right angle slump or surface cracking. The cured films have excellent tensile, elongation, flexibility, and recovery properties.

Ucar Latex 153 (55% total solids)	64.37
Pigment Paste Dispersion **Super Imperse Blue B** (50% total solids)	0.10
Calgon	0.30
Calidria Asbestos SG-210	1.16
Mica, **Mineralite 3X**	13.45
Ti Pure R-901	0.93
Clay, Kingsley	4.48
Chlorowax 70	2.09
Antimony Oxide	0.18
Dicalite SH-3	6.57

Flexol Plasticizer TCP	5.91
PMA-30	0.21
Colloid 581B (15% total solids)	0.25

Procedure:

Add the components in the order listed and mix 2 h in a 1-gal "Baker-Perkins" Sigma Blade type mixer at 79 rpm. Enclose the mixer to prevent moisture loss.

For brush or spray-gun grade, consistency let down can be made by adding extra water toward the end of the mixing cycle.

Latex Exterior Masonry Paint

Formula No. 1

	lb	gal
Rutile Titanium Dioxide	218.5	6.24
Zinc Dioxide	62.2	1.33
Talc	251.6	10.60
Diatomaceous Earth	92.2	4.80
Thixcin R	4.9	0.59
Styrene-Butadiene Copolymer Resin	54.8	6.27
Velsicol X-37	54.8	6.09
Chlorinated Paraffin	4.9	0.59
Aromatic Naphtha	204.2	28.64
Mineral Spirits	204.2	31.37
PVC (%)	57.2	

	No. 2 *(White)*	No. 3 *(Ivory)*	No. 4 *(Brown)*	No. 5 *(Gray)*	No. 6 *(Red)*
Add ingredients in order shown:					
Mineral Spirits (KB40+)	200.0	200.0	195.0	195.0	195.0
Marbon 1100T MV	71.0	71.0	83.5	83.5	83.5
Mix to clear solution					
Chlorinated Paraffin (50%)	71.0	71.0	83.5	83.5	83.5
Red Iron Oxide	–	–	–	–	120.0

Rutile Titanium Dioxide (nonchalking)	225.0	225.0	110.0	110.0	–
Zinc Oxide	50.0	50.0	50.0	50.0	25.0
Calcium Carbonate	–	–	247.0	247.0	248.0
Talc	125.0	125.0	–	–	–
Mica	50.0	50.0	75.0	75.0	70.0
Diatomaceous Silica	75.0	75.0	75.0	75.0	70.0
*Suspension Additive	4.0	4.0	4.0	4.0	4.0
Lecithin	4.0	4.0	4.0	4.0	4.0
*Denatured Alcohol	2.0	2.0	2.0	2.0	2.0

* Prewet

Mix to smooth paste and grind on 3-roll mill to fineness of 3–4 N.S.

Mineral Spirits	220.0	220.0	190.0	190.0	205.0
Lampblack in Oil	–	–	2.0	5.0	–
Yellow Iron Oxide in Oil	–	10.0	20.0	10.0	–
Total Solids (%)	61.9	61.9	66.4	65.9	63.8
PVC (%)	53.8	55.0	51.3	52.6	52.5
lb/gal	11.1	11.1	11.2	11.2	11.0
Visc. (KU)	79	79	78	78	79

Chlorinated Rubber Swimming Pool and Masonry Paint

	Formula No. 1	No. 2	No. 3
Parlon S-10	–	16.3	16.5
Parlon S-20	19.4	–	–
Chlorafin 40	7.8	6.6	6.7
Aroclor 5460	4.8	4.0	4.0
Dioctyl Sebacate (DOS)	1.5	1.3	1.3
ERL 2774	0.5	0.4	0.4
M-P-A 60 (xylene)	0.5	0.5	1.0
R900	13.6	18.9	19.2
Barytes No. 1	11.6	4.0	–
Asbestine 3X	1.8	10.3	–
Micro-Mica C-2000 (English Mica)	–	–	13.1

Xylene	30.7	30.2	30.3
Solvesso 150	7.7	7.7	7.6
Epichlorohydrin	0.07	–	–
Total Solids by Wt. (%)	62.0	62.0	62.0
PVC (%)	21.7	31.7	30.7
Total Solids by Vol. (%)	40.8	39.5	40.5
Visc. (KU)	78.0	70.0	72.0

Latex Exterior Paint for Wood and Masonry

(Yellow/Buff)

	lb	gal
Water	100	12.00
Natrosol 250 MR	2	0.17
Strodex PK-90	2	0.21
Igepal CO-630	3	0.35
Preservative	5	0.50
Drew L-475	2	0.27
Dowanol DESG	15	1.75
Ethylene Glycol	17	1.83
Rutile Titanium Dioxide (nonchalking)	120	3.42
YLO-2288B Yellow Iron Oxide	60	1.80
Vicron 45-3	200	8.85
325 Mesh W.G. Mica	25	1.06

Disperse on a high-speed disperser and let down with:

	lb	gal
Water	103	12.36
Polyco 2358	380	42.50
Water and/or **Natrosol 250 MR** (3% sol'n.) (for viscosity adjustment)	108	13.00

lb/gal	11.4
PVC (%)	41.5
Visc. (KU)	80-85

Latex Stucco and Masonry Paint

(Medium-Tone Brown)

	lb/100 gal
Rutile Titanium Dioxide (NC)	40.0
Mapico 418	80.0
Mapico XLO	40.0
Duramite	275.0
Mineralite 3X	70.0
Celite 110	70.0
Thixcin R	5.0
Soya Lecithin	10.0
Pliolite S-5A	85.0
Chlorinated Paraffin (40%)	42.5
Chlorinated Paraffin (70%)	42.5
High-Flash Naphtha	200.0
Mineral Spirits	200.0

PVC (%)	57.5
Total Solids ($)	65.5
Visc. (KU)	72

Latex Multipurpose Masonry Paint

Formula No. 1

(Gloss White–Fume Resistant)

	lb/100 gal
Pliolite S-5: Base 81	430.0
Pliolite S-5A Resin	33.0
Chlorinated Paraffin (40%)	14.5
China Wood Oil	6.5
High-Flash Naphtha	240.0
Petroleum Thinner	240.0

Total Solids (%)	50.2
PVC (%)	25.5

No. 2

(Wet Grinding–Tile Red)

	lb/100 gal
Red Iron Oxide	109.0
Clay	153.0
Pliolite S-5B	175.0
CP-40	17.5
China Wood Oil	9.0
Solvesso 100 (92 KB)	231.0
Mineral Spirits (38 KB)	231.0
Total Solids (%)	50
PVC (%)	29.5

No. 3

(Tile Red)

	lb/100 gal
Pliolite S-5: Base 7C	181.5
Base 13B	254.5
Chlorinated Paraffin (40%)	17.5
China Wood Oil	9.0
High-Flash Naphtha (92 KB)	231.0
Petroleum Thinner (38 KB)	231.0
Total Solids (%)	50.0
PVC (%)	29.5

No. 4

(Medium-Tone Red)

	lb/100 gal
Mapico 297	38
Rutile Titanium Dioxide (NC)	114
Duramite	262

Mineralite 3X	76
Celite 110	72
Thixcin R	5
Pliolite S-5A	84
Chlorinated Paraffin (40%)	42
Chlorinated Paraffin (70%)	42
Soya Lecithin	10
High-Flash Naphtha	205
Mineral Spirits	205

PVC (%)	56.3
Total Solids (%)	64.5
Visc. (KU)	78

No. 5

(Pink)

	lb/100 gal
Rutile Titanium Dioxide (NC)	141.0
Mapico Red 297	12.0
Duramite	258.0
Mineralite 3X	75.0
Celite 110	70.0
Thixcin R	5.0
Pliolite S-5A	83.0
Chlorinated Paraffin (40%)	41.5
Chlorinated Paraffin (70%)	41.5
Soya Lecithin	10.0
High-Flash Naphtha	207.0
Mineral Spirits	207.0

PVC (%)	57.4
Total Solids (%)	64.0
Visc. (KU)	77

No. 6

(Deep-Tone Brown)

	lb/100 gal
Mapico Brown 418	115.0
Duramite	288.0
Mineralite 3X	72.0
Celite 110	72.0
Thixcin R	5.0
Pliolite S-5A	82.0
Chlorinated Paraffin (40%)	41.0
Chlorinated Paraffin (70%)	41.0
Soya Lecithin	10.0
High-Flash Naphtha	204.0
Mineral Spirits	204.0

PVC (%)	56.7
Total Solids (%)	64.0
Visc. (KU)	77

No. 7

(Yellow)

	lb/100 gal
Pliolite S-5: Base 13B	187.5
Base 41A	234.5
Chlorinated Paraffin (40%)	17.0
China Wood Oil	8.5
High-Flash Naphtha	245.0
Petroleum Thinner	245.0

Total Solids (%)	47.7
PVC (%)	28

No. 8
(Olive Gray)

	lb/100 gal
Pliolite S-5: Base 81	202.5
Base 7C	0.5
Base 13B	188.0
Base 38	46.5
Base 46C	4.0
Chlorinated Paraffin (40%)	16.5
China Wood Oil	8.5
High-Flash Naphtha	232.5
Petroleum Thinner	232.5
Total Solids (%)	50
PVC (%)	31

No. 9
(Battleship Gray)

	lb/100 gal
Pliolite S-5: Base 81	206.5
Base 7C	2.0
Base 13B	188.5
Base 38	8.0
Base 46C	8.5
Chlorinated Paraffin (40%)	15.5
China Wood Oil	7.5
High-Flash Naphtha	240.5
Petroleum Thinner	240.5
Total Solids (%)	47.5
PVC (%)	31.5

No. 10
(Black–Acid and Alkali Resistant Coating)

	lb/100 gal
Pliolite S-5: Base 13B	187.0
Base 46C	140.5

Chlorinated Paraffin (40%)	16.0
China Wood Oil	8.0
High-Flash Naphtha	238.5
Petroleum Thinner	238.5
Total Solids (%)	42.4
PVC (%)	30

No. 11

(Silver Gray)

	lb/100 gal
Pliolite S-5: Base 81	223.5
Base 13B	204.0
Base 38	11.0
Base 39C	0.3
Base 46C	1.5
Chlorinated Paraffin (40%)	16.5
China Wood Oil	8.0
High-Flash Naphtha	234.0
Petroleum Thinner	234.0
Total Solids (%)	49.5
PVC (%)	31

No. 12

(Light Gray)

	lb/100 gal
Pliolite S-5: Base 81	198.5
Base 13B	161.5
Base 38	6.0
Base 39C	2.5
Base 46C	3.0
Chlorinated Paraffin (40%)	13.0
China Wood Oil	6.5

High-Flash Naphtha	251.0
Petroleum Thinner	251.0
Total Solids (%)	45.8
PVC (%)	31.5

No. 13

(Dark Green)

	lb/100 gal
Pliolite S-5: Base 82	212.0
Base 13B	229.5
Chlorinated Paraffin (40%)	16.5
China Wood Oil	8.5
High-Flash Naphtha	233.0
Petroleum Thinner	233.0
Total Solids (%)	50
PVC (%)	29.4

No. 14

(Brown)

	lb/100 gal
Pliolite S-5: Base 7C	48.5
Base 13B	110.0
Base 38	274.0
Base 46C	1.0
Chlorinated Paraffin (40%)	17.5
China Wood Oil	8.5
High-Flash Naphtha	235.0
Petroleum Thinner	235.0
Total Solids (%)	50.4
PVC (%	27.5

Acrylic Masonry Floor Paint

	Formula No. 1 (*Red*)		No. 2 (*Green*)		No. 3 (*Gray*)	
	lb	gal	lb	gal	lb	gal
Super Ad-It	4	0.48	4.0	0.48	4	0.48
Triton CF-10	2	0.24	2.0	0.24	2	0.24
Water	74	9.55	74.0	9.55	74	9.55
Ethylene Glycol	30	3.22	30.0	3.22	30	3.22
Tamol 731 (25%)	10	1.19	10.0	1.19	10	1.19
Snowflake	100	4.55	121.5	5.37	100	4.55
Silica 1160	100	4.43	122.5	5.55	100	4.43
Red Iron Oxide **RO-8097**	150	3.68	—	—	—	—
Chromium Oxide **X-1134**	—	—	70.0	1.65	—	—
Rutile Titanium Dioxide	—	—	—	—	129	3.68
Lampblack (**Cal-Ink GP-707**)	—	—	—	—	10	1.21
Cowles grind, then add:						
Rhoplex AC-61 (46%)	581	65.89	557.0	63.00	581	65.89
Diacetone Alcohol	25	2.79	25.0	2.79	25	2.79
Butyl **Cellosolve**	25	2.69	25.0	2.69	25	2.69
Colloid 600 (antifoam)	4	0.40	4.0	0.40	4	0.40
Hydroxyethyl Cellulose (3% sol'n. **WP-4400**)	60	7.16	60.0	7.16	60	7.16
PVC (%)	31		31		31	
Total Solids (%)	53.1		52.6		52.2	
Visc. (KU)	63–70		65–70		65–70	

No. 4

(Green–Low Sheen)

	lb	gal
Charge the following:		
Tamol 731 (25%)	10.0	1.19
Triton CF-10	2.0	0.24
Water	74.0	9.55
Super Ad-It	4.0	0.48
Ethylene Glycol	30.0	3.22
Snowflake	109.3	4.83
Silica 1160	122.5	5.55
Chromium Oxide **X-1134**	70.0	1.65

Grind with Cowles dissolver at 6000 rpm for 10 min. Decrease speed and add with good agitation:

Rhoplex AC-61	557.0	63.00
Diacetone Alcohol	25.0	3.19
Butyl **Cellosolve**	25.0	3.33
Colloid 600 (antifoam)	4.0	0.40
Hydroxyethyl Cellulose (3% sol'n. **WP-4400**)	60.0	7.16
Aluminum Oxide	14.5	0.54

PVC (%)	31
Total Solids (%)	52.6
Visc.–Initial (KU)	65-70

	No. 5 (High Sheen–Red)		No. 6 (High Sheen–Green)		No. 7 (High Sheen–Gray)	
	lb	gal	lb	gal	lb	gal
Charge the following:						
Tamol 731 (25%)	7.5	0.82	7.5	0.82	7.5	0.82
Triton CF-10	2.0	0.24	2.0	0.24	2.0	0.24
Colloid 600 (antifoam)	2.0	0.20	2.0	0.20	2.0	0.20
Water	107.0	12.90	51.3	6.21	75.2	9.11
Chromium Oxide **X-1134**	–	–	70.0	1.65	–	–
Red Iron Oxide **RO-8097**	158.9	3.88	–	–	–	–
Lampblack Dispersion (**Cal-Ink GP8807-B**)	–	–	–	–	30.0	2.46
Ti Pure R-900	–	–	–	–	214.2	6.12
Purecal M	62.6	2.77	105.0	4.65	–	–
Grind with Cowles dissolver at 6000 rpm for 20 min. Decrease speed and add:						
Water	–	–	36.3	4.39	24.5	2.98
Propylene Glycol	54.2	5.79	51.3	5.49	49.8	5.33
Butyl **Cellosolve**	26.4	3.51	25.0	3.33	24.4	3.24
Super Ad-It	1.0	0.12	1.0	0.12	1.0	0.12
Mix 5 min with good agitation and add:						
Rhoplex AC-61 (46%)	547.4	61.74	517.2	58.42	503.4	56.78
Aluminum Oxide	25.0	0.68	25.0	0.68	25.0	0.68
Hydroxyethyl Cellulose (3% sol'n. **WP-4400**)	68.7	8.31	65.0	7.87	63.3	7.65
PVC (%)	21.9		21.9		21.7	
Total Solids (%)	49.9		45.6		46.0	
Visc.–Initial (KU)	60–65		60–65		60–65	

No. 8

(Epoxy-Modified, Green)

	lb	gal
Super Ad-It	4.0	0.48
Triton CF-10	2.0	0.24
Water	74.0	9.55
Ethylene Glycol	30.0	3.22
Tamol 731 (25%)	10.0	1.19
Snowflake	121.5	5.37
Silica 1160	122.5	5.55
Chromium Oxide X-1134	70.0	1.65

Grind 12 min on Cowles at high speed and add with good agitation:

	lb	gal
Rhoplex AC-61 (46% solids)	501.3	56.70
Diacetone Alcohol (acetone-free)	25.0	2.79
Butyl **Cellosolve**	25.0	2.69
Colloid 600 (antifoamer)	4.0	0.40
Araldite DP-624 (50% solids)	53.5	5.35
Advacar (6% cobalt drier)	0.2	0.02
Advacar (24% lead drier)	1.1	0.11
Hydroxyethyl Cellulose (3% sol'n. **WP-4400**)	60.0	7.16

PVC (%)	31
Total Solids (%)	52.6
Visc. (KU)	65-70

No. 9

(Epoxy-Modified, Gray)

	lb	gal
Tamol 731 (25%)	7.5	0.82
Triton CF-10	2.0	0.20
Colloid 600	2.0	0.20
Water	80.4	9.73
Ti Pure R-900	228.6	6.54
Lampblack Dispersion (**Cal-Ink GP-8807B**)	30.0	2.54

Grind with Cowles dissolver at 6000 rpm for 20 min, decrease speed and add:

Water	26.1	3.16
Propylene Glycol	54.6	6.17
Super Ad-It	1.0	0.10

Mix 5 min with good agitation and add:

Rhoplex AC-61 (46%)	485.4	54.76
Araldite DP-624 (50%)	49.6	4.96
Advacar (6% cobalt drier)	0.2	0.02
Advacar (24% lead drier)	1.1	0.11
Aluminum Oxide	24.8	0.67
Butyl **Cellosolve**	24.4	3.24
Hydroxyethyl Cellulose (3% sol'n. **WP-4400**)	67.8	8.19

Total Solids (%)	48.2
PVC (%)	21.7
Visc.–Initial (KU)	60-65

Latex Interior Concrete Floor Paint

	lb/100 gal
Titanium Dioxide	150.000
Lampblack	4.000
ASP 100 or **103**	70.000
Pliolite S-5B	169.600
Chlorowax 40	17.000
China Wood Oil	8.500
Silicone Fluid	0.009*
High-Flash Naphtha	239.500
Mineral Spirits	239.500

* 0.001% of finished paint added as a 1% sol'n. in the corresponding solvent.

Visc. (KU)	67

Chapter V

ENAMELS

Enamels are characterized by gloss – a result of low pigment volume concentration. Depending on the degree of glossiness, the weathering requirements, the pigment load (either for aesthetic or mechanical reasons such as adding nonskid properties to a slippery concrete surface), and cost limitations, a myriad of trade-sales and industrial enamels exist for just about every need.

Alkyd Enamel

Formula No. 1

(High Semigloss, Acrylic Modified)

	lb	gal
Arolon 580-W-42	135.0	15.17
Cellosolve	40.0	5.20
Ethylene Glycol	15.0	1.60
KTPP	0.5	–
Surfactol 13	5.0	0.63
Cobalt Drier (6%, water dispersible)	1.0	0.12
Igepal CO-630	2.0	0.25
Triton GR-7	2.0	0.25
Titanox 2060	275.0	7.88

Disperse to 7+ N.S. grind then add

	lb	gal
Arolon 580-W-42	160.0	18.00
Thickener L	38.0	4.00
Water	38.0	4.00

Nopco NDW	4.0	0.50
Rhoplex AC707 (65% NV)	190.0	21.11
Water	175.0	21.00

PVC (%)	21.5
Visc. (KU)	75–85

No. 2

(Semigloss, with Silicone)

	lb	gal
Thixcin R	2.0	0.3
Soya Lecithin	10.0	1.2
Varkyd 333-60	477.0	60.0
Mineral Spirits	130.0	20.0
Ranc	170.0	5.0
Desertalc 55	95.0	4.1
Atomite	75.0	3.3
Micromica C-3000	50.0	2.1
Lampblack Pigment	3.1	0.3
Ethyl Benzene	14.5	2.0
Pebble mill:		
Xylol	14.5	2.0
Cobalt Naphthenate (6%)	1.0	0.1
Manganese Naphthenate (6%)	1.0	0.1
Calcium Naphthenate (4%)	4.0	0.5
Lead Naphthenate (24%)	5.0	0.5
Antiskinning Agent	1.0	0.1
Dow Corning #1 Additive	3.1	0.4

PVC (%)	31.5
Vehicle NV by Wt. (%)	28.3
Total NV by Wt. (%)	65.5
lb/gal	10.34
Visc. (KU)	68

No. 3

(Gloss)

	lb	gal
Varkyd 552-50 or **1554-50**	190.0	25.0
Nuosperse 657	5.0	0.6
MPA-60X	7.0	0.8
Mineral Spirits	52.0	8.0
Amsco Solvent B	21.5	3.0
Titanium Dioxide 2061	250.0	7.3

Sand mill or pebble mill:

	lb	gal
Varkyd 552-50 or **1554-50**	448.0	59.0
Cobalt Naphthenate (6%)	3.0	0.4
Lead Naphthenate (24%)	7.0	0.7
Exkin #2	1.0	0.1

PVC (%)	17.0
Vehicle NV by Wt. (%)	44.9
Total NV by Wt. (%)	58.5
lb/gal	9.38
Visc. (KU)	74

No. 4

(Mix-In-Type, Gloss)

	lb	gal
Rutile Titanium Dioxide	275.0	7.95
Surfex MM	183.5	8.30
Soya Lecithin	3.1	0.36
MPA 60	4.0	0.59
Alkyd Resin Solids (70%)	278.0	35.00
Mineral Spirits	43.8	6.73
Lead Naphthenate (24%)	2.4	0.25

Mix on Cowles dissolver at 5100 rpm until temperature of 60 C is obtained. Add the following:

Alkyd Resin Solids (70%)	176.0	22.10
Mineral Spirits	119.0	18.15
Cobalt Naphthenate (6%)	3.6	0.45
Antiskinning Agent	1.0	0.12

No. 5

(Gloss, Odorless)

	lb	gal
Ti Pure R-900	300	8.60
Kadox 515 (lead-free)	20	0.42
Aroplaz 1241-Mo-65 (65% NV)	550	70.50
Odorless Mineral Spirits	118	18.60
Cobalt Naphthenate (6%)	3	0.38
Calcium Naphthenate (4%)	8	1.00
Antiskinning Agent	1	0.13

Visc. (KU)	83
PVC (%)	18.4
lb/gal	10.00

No. 6

(Semigloss, Isophthalic)

	lb	gal
Zopaque R-55	300.0	8.45
Snowflake	150.0	6.63
IB-1550 (50% NV)	450.0	59.30
Mineral Spirits	88.0	13.50
Cobalt Naphthenate (6%)	1.8	0.22
Lead Naphthenate (24%)	4.6	0.48
Antiskinning Agent	1.5	0.18

Visc. (KU)	79

No. 7

(Semigloss, Isophthalic Alkyd)

	lb/100 gal
IB-1550	492.00
N Lecithin	3.00
R-900	250.00
Atomite	200.00
Thixatrol ST	5.00
Mineral Spirits	113.00
Zirco Naphthenate (6%)	4.00
Cobalt Naphthenate (6%)	2.50
Antiskinning Agent	1.00

Procedure:

High-speed disc impeller (Cowles) at approximately 150 F.

PVC (%)	38.0
lb/gal	10.71
Visc. (KU)	85 ± 2
Total Vehicle Solids (%)	40.65

No. 8

(Undercoater)

	lb	gal
Beckosol P-471-50	385.0	51.00
Yelkin TSS	4.0	0.45
Thixatrol ST	3-5.0	0.34
Zopaque R-77	200.0	5.86
Drikalite	425.0	18.83

Disperse using high speed disperser adding in above order.

Mineral Spirits	148.0	22.75
Lead Tallate (24%)	4.1	0.44
Cobalt Tallate (6%)	1.7	0.21
Antiskinning Agent	1.0	0.12

lb/gal	11.7
Total NV (%)	70.6
PVC (%)	53.4
Visc. (KU)	92

Latex Semigloss White or Light Tint Base Enamel

	Formula No. 1	No. 2
Propylene Glycol	62.7	62.7
Tamol 731	13.5	13.5
Nopco NDW	1.9	1.9
Rutile Titanium Dioxide (enamel grade)	250.0	225.0
Barytes	50.0	–
White NC Clay	–	25.0
Celite 499	–	12.5

Disperse with a disc disperser at high speed. Then add at low speed:

Propylene Glycol	105.8	105.8
Cellosize WP-4400 (2.5% sol'n.)	43.6	43.6
Nopco NDW	1.9	1.9
Triton GR-7	1.9	1.9
Rhoplex AC-22	412.2	412.2
Rhoplex AC-61	99.3	99.3
Super Ad-It	0.9	0.9
Water	58.3	58.3
Visc. (KU)	80	82

Latex Interior Flat Enamel

	lb	gal
Dispersion:		
Water (as needed)	261	31.5
Bentone LT	3	0.2

Grind 5–10 min and add:

Bioxide	1	0.1
Nopco NDW	1	0.1
Ammonia (28% sol'n.)	2	0.2

Strodex PK-90	1	0.1
Strodex SEK-50	2	0.2
Titanox 2030	275	8.6
Optiwhite	75	4.1
Imsil A-10	25	1.1

Grind at high speed:
Let down:

Ethylene Glycol	30	3.2
Igepal CO-630	4	0.5
Nopco NDW	1	0.1
Rhoplex AC-490	442	50.0

Total Vehicle Solids by Vol. (gal)	21.2
Total Solids by Vol. (gal)	35.0
Total Solids by Wt. (%)	52.0
PVC (%)	39.4

Polyester High Gloss Metallized Enamel

	lb/100 gal
Aluminum Paste 222	10.1
Cyplex 1473-5 (65%)	306.9
Cymel 248-8	155.4
SR-112 (50%)	5.7
Dow Corning Paint Additive No. 1	1.4
Ethylene Glycol Monoethyl Ether Acetate	15.3
Isophorone	30.5
Methyl Isobutyl Ketone	71.3
Xylol	86.6
Toluene	115.8
Butanol	5.1

Note:

Binder: **Cyplex 1473-5/Cymel 248-8** = 70/30, solids basis. No catalyst required.

Total NV by Wt. (%)	36.7
Total NV by Vol. (%)	28.6
Visc. (s–Ford Cup No. 4)	18

Polyurethane Gloss White Enamel

		lb	gal
A	**Ti Pure R-900**	180.0	5.30
	M-P-A 60 (xylene)	3.5	0.48
	Cellosolve Acetate	10.1	1.25
	Stepan Alkyd 4209	72.0	8.37
Three roll mill grind, 3 passes			
	Stepan Alkyd 4209	364.0	42.38
	SF-69 (2% in **Cellosolve** acetate)	2.5	0.31
	Chevron #3	22.2	3.00
	Ethyl Benzene (99%)	16.3	2.25
	Cellosolve Acetate	281.0	34.66
B	**Nacconate H12**	20.1	2.22
	Ethyl Benzene (99%)	20.1	2.78

Total Solids by Wt. (%)	42.3
Visc. (s–#4 Ford at 25 C)	107.6

Acrylic Aluminum Enamel

(Alkyd Modified, Fast-Dry)

Roller mill grind:

Rutile Titanium Dioxide	65.0
Amberlac 292X (48%)	35.0

Let down:

Amberlac 292X (48%)	23.5
Acryloid B-66 (40%)	170.5
Xylene	40.0

Total Solids (%)	53.5
P/B Ratio	4:6
Visc. (s–No. 4 Ford Cup)	≈ 65

Alkyd Floor Enamel

	lb	gal
Methocel 4000 cps (2%)	180	21.60
Tamol 731 (25% sol'n.)	5	0.60
Ti Pure R-900	100	2.86
Natural Barium Sulfate	100	2.70
ASP-100	75	3.49
Magnesium Silicate	25	1.16
Butrol	1	0.10
Nopco JMK	2	0.24
High speed disperser		
Water	160	19.20
Dow Latex 762-W	208	24.80
Arolon 110-W-45	220	25.70
Cobalt Drier (6%)	2	0.25
Carbon Black Tint	2	0.20
PVC (%)	30.2	
Visc. (KU)	65-70	
Total Solids (%)	46.92	

Acrylic Color-Retentive, General-Purpose Enamel

	lb	gal
Roller mill grind:		
Titanium Dioxide	142.5	4.4
Acryloid B-82 (50% solids)	106.8	12.7
Xylene	44.5	6.3
Mix with:		
Acryloid B-82 (50% solids)	320.8	38.2
Xylene	196.2	27.1
Toluene	80.2	11.3
lb/gal	8.9	
Total Solids (%)	40.0	
P/B Ratio	4:6	

Acrylic Gloss White Enamel

Formula No. 1

Ball mill grind:

Zopaque RCL-9	480.0
Acrysol WS-68 (38%)	126.3
Water	149.4
Butyl **Cellosolve**	21.4
n-Butanol	20.0
DMAE	2.9

Let down:

Ball mill grind	326.4
Acrysol WS-68 (38%)	567.0
Water	48.2
n-Butanol	6.9
L-5310 (30% in butyl **Cellosolve**)	1.9
DMAE (10% in water)	5.1
Cymel 303	58.7
PTSA (10%)	5.8

Total Solids (%)	48.0
P/B Ratio: 40:60	
pH	7.7
Visc. (s–No. 4 Ford Cup)	23

No. 2

Sand mill (30 min):

Cymel 303	63.0
Tamol 731 (25%)	4.2
DMAE (20% in water)	5.0
Butyl **Cellosolve**	10.4
Isopropanol	5.6
Water	101.8
Zopaque RCL-9	210.0

Let down:

Sand mill grind		373.0
Acrysol WS-68 (38%)		618.4
Water		4.0
DMAE (20% in water)		10.8
L-5310 (30% in butyl **Cellosolve**)		2.0
PTSA (10%)		5.8
n-Butanol		6.0
Total Solids (%)	48.0	
P/B Ratio	40:60	
pH	7.6	
Visc. (s–No. 4 Ford Cup)	25	

No. 3

(Premium Quality)

		lb	gal
Propylene Glycol		80.0	9.28
Tamol 731 (25%)		5.0	0.55
Nopco NDW		2.0	0.26
Rutile Titanium Dioxide		250.0	7.30

Grind the above in a high-speed mill (Cowles) for 20 min and let down at a slower speed as follows:

		lb	gal
Water		50.0	6.00
Rhoplex AC-25 (46.5%)		537.0	60.60
Propylene Glycol	Premix	40.0	4.60
Texanol		15.8	2.00
Water	Premix	16.6	2.00
Dowicil 75		1.3	–
Nopco NDW		4.0	0.52
Water/2% **Natrosol 250 MR** sol'n.		57.4	6.89
PVC (%)		22	
Total Solids by Vol. (%)		33.2	
pH–Initial		8.5-9.0	
Visc.–Initial (KU)		72-75	

Acrylic Gloss Black Enamel

Sand mill (30 min)

Cymel 303	135.6
Tamol 731 (25%)	1.2
Butyl **Cellosolve**	21.0
Isopropanol	21.0
Water	84.0
DMAE (20% in water)	1.5
Raven 1035	35.7

Let down:

Sand mill grind	150.3
Acrysol WS-68 (38%)	715.5
PTSA (10%)	6.8
L-5310 (30% in butyl **Cellosolve**)	2.2
n-Butanol	5.2

Total Solids (%)	40.7
P/B Ratio	5/95
pH	7.6
Visc. (s–No. 4 Ford Cup)	23

Acrylic Tintable Eggshell Enamel

	lb	gal
Propylene Glycol	51.8	6.00
Water	50.0	6.00
Natrosol 250 GR	2.0	–
Tamol 731 (25%)	7.0	0.76
Nopco NDW	2.0	0.26
Rutile Titanium Dioxide	250.0	7.30
Silica (small particle, amorphous)	75.0	3.40

Grind the above in a high-speed mill (Cowles) for 20 min and let down at a slower speed as follows:

Water	50.0	6.00
Nopco NDW	4.0	0.52
Rhoplex AC-25 (46.5%)	450.0	50.60

Texanol	Premix	12.6	1.59
Propylene Glycol		17.3	2.00
Water	Premix	24.9	3.00
Dowicil 75		1.3	–
Triton N-57		4.0	0.44
Triton X-114		4.0	0.44
Triton X-207		2.0	0.22
Acrysol G-110	Premix	10.0	1.13
Water		25.0	3.00
Water/4% **Natrosol 250 GR** sol'n.		61.1	7.34

PVC (%)	33.2
Total Solids by Vol (%)	32.3
pH–Initial	8.5-9.0
Visc.–Initial (KU)	72-75

Acrylic Water-Reducible Enamel

Formula No. 1

(Yellow)

	lb	gal
1. **Medium Chrome Yellow 40-4500**	150.6	3.12
2. **XC-4011** (75%)	124.7	14.67
3. Dimethylamino Ethanol	9.7	1.31

Disperse on 3-roll mill:

4. **XC-4011** (75%)	124.9	14.69
5. **Cymel 303**	46.9	4.69
6. **Catalyst 4040**	1.2	0.16
7. Dimethylamino Ethanol	9.9	1.34
8. Demineralized Water	499.9	60.04

Procedure:

Premix items 6 and 7. Add to a suitable vessel followed by items 4 and 5. Mix under high speed mixer using slow speed. Disperse item 1 in items 2 and 3 on 3-roll mill. Pull mill base into vessel with items 4-7. Mix until uniform. Continue mixing while slowly adding item 8.

Total NV by Wt. (%)	39.8
Total NV by Vol. (%)	28.0
PVC (%)	11.1
Visc. (s–Ford Cup No. 4)	70

No. 2

(Red)

	lb	gal
1. **Toluidine Red 20-3300**	54.0	4.32
2. **XC-4011** (75%)	111.6	13.13
3. Dimethylamino Ethanol	10.6	1.43
4. Butyl **Cellosolve**	18.0	2.39
5. **Cymel 303**	42.0	4.20
6. **BYK-P-104-S**	1.7	0.21

Disperse above on Cowles dissolver:

7. **XC-4011** (75%)	112.1	13.19
8. **Catalyst 4040**	1.1	0.15
9. Dimethylamino Ethanol	9.9	1.34
10. Demineralized Water	496.5	59.63

Procedure:

In a suitable vessel under Cowles, premix items 2-6. Add item 1 and disperse using high speed. Reduce to slow speed. Premix items 8 and 9. Add to vessel followed by item 7. Mix until uniform. Finally add item 10 slowly while continuing to mix using slow speed.

Total NV by Wt. (%)	31.0
Total NV by Vol. (%)	26.8
PVC (%)	16.1
Visc. (s–Ford Cup No. 4)	70

No. 3

(Blue)

	lb	gal
1. Cyan Blue GT-55-3295	33.6	2.38
2. **Unitane OR-650**	10.6	0.32
3. **XC-4011** (75%)	105.1	12.37
4. Dimethylamino Ethanol	9.9	1.34
5. **BYK-P-104-S**	2.0	0.26
6. Demineralized Water	229.3	27.52

Disperse on sand mill:

	lb	gal
7. **XC-4011** (75%)	108.0	12.70
8. **Cymel 303**	40.3	4.03
9. **Catalyst 4040**	1.0	0.14
10. Dimethylamino Ethanol	9.8	1.32
11. Demineralized Water	313.4	37.62

Procedure:

Step 1 – Premix items 9 and 10. Add to suitable vessel with items 7 and 8. Mix until uniform. *Step 2* – Premix items 1-6. Pass through sand mill and pull into vessel containing items 7-10. Place vessel under high speed mixer and add item 11 slowly while mixing under slow speed.

Total NV by Wt. (%)	28.6
Total NV by Vol. (%)	24.2
PVC (%)	11.2
Visc. (s–Ford Cup No. 4)	70

No. 4

(Black)

	lb	gal
1. Carbon Black	12.0	0.83
2. **XC-4011** (75%)	118.8	13.97
3. Dimethylamino Ethanol	11.2	1.52
4. **BYK**	2.3	0.29
5. Demineralized Water	267.7	32.13

Disperse above on sand mill:

6. **XC-4011** (75%)	121.7	14.32
7. **Cymel 303**	45.5	4.55
8. **Catalyst 4040**	1.2	0.15
9. Dimethylamino Ethanol	11.2	1.52
10. Demineralized Water	255.8	30.72

Procedure:

Premix items 8 and 9. Add to suitable vessel with items 6 and 7. Mix until uniform. Then premix items 1-5. Pass through sand mill and pull into vessel containing items 6-9. Place vessel under high speed mixer and add item 10 slowly while mixing under slow speed.

Total NV by Wt. (%)	28.4
Total NV by Vol. (%)	25.1
PVC (%)	3.3
Visc. (s–Ford Cup No. 4)	70

No. 5

(White, High Gloss)

	lb	gal
1. Rutile Titanium Dioxide	182.7	5.49
2. **XC-4011** (75%)	121.6	14.31
3. Dimethylamino Ethanol	9.5	1.29
4. **XC-4011** (75%)	121.8	14.33
5. **Cymel 303**	45.8	4.58
6. **Catalyst 4040**	1.2	0.16
7. Dimethylamino Ethanol	9.6	1.30
8. Demineralized Water	484.4	58.54

Procedure:

Disperse item 1 in items 2 and 3. Add mill base to suitable vessel under a high speed mixer. Add items 4 and 5 to vessel. Premix items 6 and 7 and add to vessel. Mix until uniform using slow speed. Add item 8 slowly while continuing to mix.

Total NV by Wt. (%)	42.0
Total NV by Vol. (%)	29.7
PVC (%)	18.7
Visc. (s–Ford Cup No. 4)	70

No. 6

(White, High Gloss)

	lb	gal
1. Rutile Titanium Dioxide	182.9	5.49
2. **XC-4011** (75%)	121.8	14.33
3. Dimethylamino Ethanol	9.4	1.28
Disperse on 3-roll mill:		
4. **XC-4011**	90.2	10.63
5. **Cymel 1123**	68.7	7.08
6. **Catalyst 4040**	3.5	0.47
7. Dimethylamino Ethanol	7.4	1.00
8. Demineralized Water	496.9	59.72

Procedure:

Disperse item 1 in items 2 and 3. Premix items 6 and 7. Add to suitable vessel under high speed mixer. Add items 4 and 5. Mix until uniform. Add mill base and remix until uniform. Finally add item 8 slowly while continuing to mix using slow speed.

Total NV by Wt. (%)	42.0
Total NV by Vol. (%)	29.8
PVC (%)	18.4
Visc. (s–Ford Cup No. 4)	70.0

No. 7

(White, High Gloss)

	lb	gal
1. Rutile Titanium Dioxide	190.8	5.73
2. **XC-4011**	127.0	14.95
3. Dimethylamino Ethanol	11.6	1.58

Disperse above on 3-roll mill:

4. **XC-4011**	64.1	7.55
5. **Cymel 370**	108.6	11.07
6. Dimethylamino Ethanol	5.8	0.79
7. Butanol	25.6	3.80
8. Demineralized Water	454.3	54.53

Procedure:

Disperse item 1 in items 2 and 3 using a 3-roll mill. In a suitable vessel under a high speed mixer, add items 4-7. Premix and add mill base. Continue mixing until uniform. Finally add item 8 slowly while continuing to mix.

Total NV by Wt. (%)	43.5
Total NV by Vol. (%)	30.3
PVC (%)	18.9
Visc. (s–Ford Cup No. 4)	45-50

No. 8

(White, High Solids)

	lb	gal
1. Rutile Titanium Dioxide	410.3	12.32
2. **XC-4011** (75%)	273.1	32.13
3. Dimethylamino Ethanol	21.2	2.87

Disperse on 3-roll mill

4. **XC-4011** (75%)	273.5	32.18
5. **Cymel 303**	102.8	10.28
6. **Catalyst 4040**	2.6	0.35
7. Butanol	46.7	6.90
8. Dimethylamino Ethanol	21.9	2.97

Procedure:

Disperse item 1 in items 2 and 3. Premix items 6 and 8. Add to suitable vessel under high speed mixer, followed by items 4, 5, and 7. Mix until uniform using slow speed. Add mill base and continue mixing until uniform. If reduction is desired, add water or organic solvents while continuing to mix.

Total NV by Wt. (%)	80.2
Total NV by Vol. (%)	66.8
PVC (%)	18.5
Visc. (Brookfield, poises)	536

No. 9

(Gray, Dipping)

	lb	gal
Rutile Titanium Dioxide	187.1	5.62
Mapico Lemon Yellow Oxide	17.9	0.53
Super Jet Lampblack	4.9	0.33
Red Iron Oxide	3.6	0.09
Aerosil R-972	12.0	0.65
XC-4012	142.7	16.40
Dimethylamino Ethanol	5.3	0.72
Demineralized Water	111.5	13.38
Cymel 303	54.0	5.40
Dapro DF-911	7.9	1.10

Disperse above on sand mill

	lb	gal
XC-4012	278.6	32.02
Catalyst 4040	1.3	0.17
Dimethylamino Ethanol	9.6	1.30
Demineralized Water	184.9	22.20

Acrylic Latex Enamel

Formula No. 1

(Battleship Gray)

	lb/100 gal
Rutile Titanium Dioxide (NC)	150.000
Lampblack	4.000
ASP 103	70.000

Pliolite AC	170.000
Chlorinated Paraffin 40	17.000
China Wood Oil	8.500
High Flash Naphtha	240.000
Mineral Spirits	240.000
Add:	
Viscasil (60,000 cps)	0.009*
PVC (%)	26.0
Total Solids by Wt. (%)	46.6

* 0.001 percent of finished paint–added as 1 percent master solution in solvent.

No. 2

(Red)

	lb/100 gal
Red Iron Oxide	118.0
ASP 103	105.0
Pliolite AC	166.0
Chlorinated Paraffin 40	17.0
China Wood Oil	8.5
High Flash Naphtha	243.0
Mineral Spirits	243.0
PVC (%)	25.8
Total Solids by Wt. (%)	45.9

No. 3

(Chromium Oxide Green)

	lb/100 gal
Chromium Oxide	164.0
ASP 103	88.0
Pliolite AC	173.0
Chlorowax 40	17.0

China Wood Oil		8.5
High Flash Naphtha		240.0
Mineral Spirits		240.0
PVC (%)	25.7	
Total Solids by Wt. (%)	48.4	

Acrylic-Urethane White Enamel

A	Titanium Dioxide	283.1
	Acryloid AU-568	113.9
	n-Butyl Acetate (Polyurethane grade-PUG)	97.7
	2-Ethoxyethyl Acetate (PUG)	95.6

Grind with equal weight of sand to desired dispersion then let down with:

	Acryloid AU-568	181.6
B	**Desmodur N-75**	232.3
	n-Butyl Acetate (PUG)	28.1
	2-Ethoxyethyl Acetate (PUG)	27.5

To the mixture of Part A and Part B, add 0.4 parts by weight (0.1% based on total resin solids) of

SF-1023 Silicone		0.4
Total Solids by Wt. (%)	67	
Total Solids by Vol. (%)	55	
Visc. (s–No. 4 Ford Cup)	20	
P/B Ratio	40:60	

Alkyd White Enamel

Formula No. 1

(Gloss)

	lb	gal
Titanox RA-45	300.0	8.58
AZO-ZZZ-77	15.0	0.32
Syntex 70 (65% NV)	515.0	65.00

Mineral Spirits	150.0	22.70
Lead Naphthenate (24%)	2.7	0.28
Cobalt Naphthenate (6%)	1.7	0.21

PVC (%)	19.50
lb/gal	10.25
Visc. (KU)	74

No. 2

(Gloss, Force Dry)

	lb	gal
Arolon 585-W-43	127.3	14.30
Ti Pure R-902HG	211.8	6.35
Butyl **Cellosolve**	20.1	2.68
Triton X-405	5.9	0.65
Cobalt Naphthenate (6%)	2.2	0.27
Lead Naphthenate (24%)	0.4	0.04
Water	47.0	5.64

Pebble mill to 7½ Hegman

Arolon 585-W-43	378.5	42.53
Arolon 376-WA8-50	141.6	16.27
Water	94.0	11.28

NV (%)	48.9
P/B Ratio	0.73:1.00
Visc. (s–No. 4 Ford Cup)	80–120
pH	7.0–7.5

Procedure:

Charge the **Arolon 585** in a mix tank and add **Arolon 376**/water. Mix slowly with agitation. After this conditioning step it may be combined with the pigment grind. 4.7 lb of 10% **L-5310** silicone sharpens the gloss.

No. 3

(For Rollercoat)

	lb	gal
Ti Pure R-902 HG	282.1	8.47
Arolon 465-WA8-70	95.9	10.65
Water	145.6	17.51
Pebble mill to 7½ Hegman and add:		
Arolon 465-WA8-70	265.3	29.47
Cymel 301	64.8	6.48
Texanol	63.7	8.06
Dimethylethanolamine	1.7	0.22
Water	159.1	19.14
Total NV (%)	55.6	
P/B Ratio by Wt.	0.89 : 1.00	
Visc. (s–No. 4 Ford Cup)	80–100	

No. 4

(Air Dry)

	lb	gal
High speed disperser grind:		
R-900	170.0	4.86
Arolon 363-WA8-50	120.0	13.95
Mineral Spirits (odorless)	3.4	0.52
High speed disperser–20 min		
Arolon 363-WA8-50	305.0	35.47
Lead Naphthenate (24%)	0.4	0.04
Cobalt Naphthenate (6%)	1.8	0.22
Water	375.0	45.07
Triethylamine	1.0	0.17
lb/gal	9.74	
Total NV (%)	39.3	
P/B Ratio	0.8	

No. 5

(With Silicone, Air-Drying)

	lb	gal
Pigment grind in sand mill:		
Chempol 11-3620	406.97	54.26
Titanium Dioxide	298.44	8.73
Lexinol AC-1	10.17	1.26
M-P-A 60 (xylene)	5.09	0.70
Let down:		
Chempol 11-3620 (50% NV)	67.97	9.06
R-641 Silicone Resin (60% NV)	169.57	18.80
Xylene	33.91	4.68
Varsol No. 1	9.87	1.50
Cobalt Naphthenate (6%)	3.39	0.42
Manganese Naphthenate (6%)	1.36	0.17
Calcium Naphthenate (5%)	1.36	0.17
Antiskinning Agent	2.03	0.25

Total NV (%)	63.2
PVC (%)	18.3
P/B Ratio	83.3:100
Visc. (Brookfield, cps)	180

No. 6

(Isophthalic)

	lb/100 gal
IB-1550	568.00
RA-45	300.00
Zinc Oxide	7.50
Thixcin R	7.00
Amsco 460	45.00
Mineral Spirits	52.00
Zirco Naphthenate (6%)	4.50
Cobalt Naphthenate (6%)	3.00
Antiskinning Agent	0.75

Procedure:
Pebble mill or roller mill at approximately 110 F.

PVC (%)	23.5
lb/gal	9.85
Visc. (KU)	85 ± 2
Vehicle Solids (%)	43.0

	No. 7	No. 8	No. 9	No. 10
Duraplex D 65 A-70	–	285.0	237.0	187.0
Aroplaz 1085 M-50	455.0	–	–	–
Velsicol XL-30	–	–	34.0	6.80
Mineral Spirits 36 KB	70.0	152.0	–	–
L.O. 460 Solvent, 31 KB	–	–	166.0	158.0
Nuodex Cobalt (6%)	2.0	2.0	2.0	2.0
Nuodex Lead (24%)	5.0	5.0	5.0	5.0
Exkin No. 2	1.0	1.0	1.0	1.0
Total Solids (%)	59.3	65.5	65.5	67.5
Total Vehicle Solids (%)	43.2	53.0	53.0	54.8

	No. 11 *(Four-Hour, Medium)*	No. 12 *(Four-Hour, Long)*
Ti Pure R-900	250.0	250.0
Nuvis HS	7.0	7.0
Nuact Paste	3.5	3.5
Nuodex Calcium (4%)	3.5	3.5
Aroplaz 1241 ML-70	–	205.0
Aroplaz 1085 M-50	130.0	–

No. 13

(With Urea-Formaldehyde)

Unitane OR 342	250
Oxidizing Alkyd Resin (55% solids in xylol)	382

Beetle XB-1032 Resin (50%)	138
Xylol	206

P/B Ratio	90:100
Visc. (s–No. 4 Ford Cup)	50–60

Alkyd Red Enamel

Formula No. 1

(Air-Dry)

	lb	gal
Arolon X322 (50% NV)–Ammonia Neutralization	121.48	13.82
Garnet Toner	11.80	0.92
Triethylamine	1.33	0.22
Water	108.62	13.04
Pebble mill 19 h		
Molybdate Orange	32.88	0.71
Pebble mill 5 h		
Arolon X322 (50% NV)–Ammonia Neutralization	341.27	36.46
Lead Naphthenate (24%)	3.65	0.39
Cobalt Naphthenate (6%)	1.50	0.19
Manganese Naphthenate (6%)	1.52	0.19
n-Butanol	33.21	4.92
Flow Additive	4.43	0.54
Water	238.24	28.60

P/B Ratio	0.2
Total Solids (%)	31.0
Visc. (s–No. 4 Ford Cup)	40–50

No. 2

(Dip)

	lb	gal
RT-3140	71.8	5.90
Arolon 580-W-42	182.2	20.49
Ethoxy Ethanol	55.2	7.12

Lecithin #55 (water-dispersible)	4.4	0.52
Deionized Water	49.7	5.98

Steel ball mill to 7½ Hegman

Arolon 580-W-42	433.0	48.71
Resimene X-720 (80%)	41.1	4.11
Propasol B	47.9	6.54
L-5310 (10%)	5.2	0.63
NV (%)	40.9	
P/B Ratio	0.25 : 1.00	

No. 3

	lb
Monastral Red	8.0
BON Red Dark	35.0
Molybdate Orange-**Valencia 40-8300**	40.0
Nonoxidizing Alkyd (60% solids)	464.0
Cymel 255-10	186.0
Xylol	133.0
Butanol	6.0
P/B Ratio	22 : 100
Visc. (s—No. 4 Ford Cup)	50–60

Alkyd Blue Enamel

Formula No.1

Phthalocyanine Blue 55-3750	19
Unitane OR540	98
Oxidizing Alkyd (55% solids)	402
Cymel 255-10	147
Xylol	223
P/B Ratio	40 : 100

No. 2

	lb	gal
Arolon X322 (50% triethylamine neutralized)	58.89	6.70
Ti Pure R-900	68.30	2.00
BP-3660	11.20	0.83
Triethylamine	0.91	0.15
Water	61.64	7.40

Pebble mill to 7½ Hegman

Arolon X322 (50% triethylamine neutralized)	334.28	38.03
Lead Naphthenate	3.37	0.36
Cobalt Naphthenate	1.34	0.17
Manganese Naphthenate	0.64	0.08
L-5310 (10%)	4.10	0.50
Water	364.69	43.78

P/B Ratio	0.4:1.0
NV (%)	30.4
pH @ 25 C	7.5-8.0
Visc. (s—No. 4 Ford Cup)	50-60

Alkyd Black Enamel

Formula No. 1

	lb	gal
Arolon X322 (50% NV ammonia neutralized)	59.24	6.74
Beaded Carbon Black	9.90	0.66
Triethylamine	1.88	0.31
Water	58.31	7.00

Steel ball mill to 7½ grind

Arolon X322 (50% NV ammonia neutralized)	335.43	38.16
Lead Naphthenate (24%)	3.28	0.35
Cobalt Naphthenate (6%)	1.35	0.18
Manganese Naphthenate (6%)	1.44	0.18

n-Butanol	29.63	4.39
L-5310 (10%)	3.94	0.48
Water	346.11	41.55

P/B Ratio	0.05
Solids (%)	24.4
Visc. (s—No. 4 Ford Cup)	50-60

No. 2

Carbon Black Mark II	33
Nonoxidizing Alkyd (60% solids)	391
Cymel 255-10	144
Xylol	250

P/B Ratio	10:100

No.3

(Gloss)

	lb	gal
Arolon 525-W-37	103.9	11.94
Propasol B	16.6	2.27
Union Carbon R	12.5	0.83

Steel ball mill to Hegman 7½ and add

	lb	gal
Arolon 525-W-37	537.5	61.57
Propasol B	59.5	8.13
Manganese Naphthenate (6%)	1.8	0.22
L-5310 (10%)	5.0	0.59
Deionized Water	120.2	14.45

Total NV (/%)	29.3
pH as is	8.3-9.0
Visc. (s—No. 4 Ford Cup)	30-50

Alkyd Interior Orange Enamel

	lb	gal
Arolon 969 (30%)–Ammonia Neutralization	122.80	14.23
Imperial X-2806	65.60	11.31
Water	12.25	1.47

Pebble mill 24 h

	lb	gal
Arolon 969 (30%)–Ammonia Neutralization	494.07	57.25
n-Butanol	27.82	4.14
Manganese Naphthenate (6%)	0.56	0.07
Cobalt Naphthenate (6%)	0.55	0.07
L-5310 (10%)	3.68	0.45
Water	91.71	11.01

Total NV	31.1
P/B Ratio	1:2.8
lb/gal	8.19
Visc. (s–No. 4 Ford Cup)	40–50

Alkyd Interior Toluidine Red Enamel

	lb	gal
RT-3140	91.93	7.56
Arolon 585-W-43	233.09	26.19
Arolon 580-W-42	72.71	8.17
Isopropanol	15.26	2.33
Dispersing Agent	6.16	0.77
Tamol 731 (25%)	3.13	0.34
Cobalt Naphthenate	2.69	0.34
Lead Naphthenate	0.56	0.06
Water	45.82	5.50

Steel ball mill 24–48 h

	lb	gal
Arolon 585-W-43	422.66	47.49
Propasol Solvent B	9.15	1.25

P/B Ratio	0.3
NV Solids (%)	44.1
Visc. (s–No. 4 Ford Cup)	50–60

Alkyd Chrome Yellow Enamel

(With Urea Formaldehyde)

Cyanamid 40-4480		150
Oxidizing Alkyd Resin (50% solids in xylol and mineral spirits)		435
Beetle XB-1032		184
Xylol		153
P/B Ratio	49:100	
Visc. (s–No. 4 Ford Cup)	50–60	

Epoxy Enamel

Formula No. 1

(White Tinting)

	lb	gal
Epon 1001-X-75	474.1	52.10
R-900	323.9	9.50
Talc No. 399	35.4	1.60
Multifex MM	35.5	1.60
Nuosperse 657	6.3	0.80
M-P-A 60 (xylene)	19.0	2.60
PC-1344 (60% NV xylene)	2.6	0.30
Advance Antiflood Agent	4.1	0.50

Two passes over 3-roll mill
Let down:

	lb	gal
Beetle 216-8	9.4	1.10
Solvesso 150	118.5	15.90
Diacetone Alcohol	27.6	3.50
Butyl Oxitol	39.5	5.30
Oxitol	39.5	5.10

Curing Agent Component:

	lb	gal
Epon Curing Agent **VI-KX-60**	586.6	75.20
Solvesso 150	118.3	15.90
DMP-30	7.3	0.90

Oxitol	16.3	2.10
Butyl Oxitol	33.8	4.50
Advance Antiflood Agent	4.1	0.50

No. 2

(White, Semigloss)

	lb	gal
Pigment Dispersion:		
Aroflint 252	270.0	28.80
Ethylene Glycol Monoethyl Ether Acetate	16.2	2.00
Rutile Titanium Dioxide	300.0	8.58
Calcium Carbonate	200.0	8.90
Magnesium Silicate	175.0	7.35
Thixatrol ST	1.0	0.16
Grind on high speed disperser to 140 F		
Mineral Spirits (exempt)	31.6	4.88
Aroflint 252	57.2	6.00
Clear Activator:		
Aroflint 606	160.0	19.05
Mineral Spirits (exempt)	71.5	11.00
Ethylene Glycol Monoethyl Ether Acetate	26.6	3.28
Enamel Formula:		
Pigment Dispersion	1051.1	66.67
Clear Activator	258.1	33.33

Epoxy-Polyamide Enamel

Formula No. 1

(White Universal, Tint-Base)

	lb	gal
Base Component:		
Vanoxy 201-X-75	233.3	25.64
Rayox R-88	270.0	7.85
IAF X-2280	6.0	0.20

M-P-A (xylene)	4.0	0.55
Soya Lecithin	1.3	0.15
Beetle 216-8	9.0	1.06
Vansolve EE	11.2	1.44
Vansolve EB	58.0	7.72
Solvesso 150	40.0	5.39

Curing Agent Component:

Van-Amid 300-KX-60	248.4	31.85
DMP-30	4.6	0.57
Vansolve EE	52.9	6.83
Vansolve EB	15.2	2.02
Solvesso 150	64.8	8.73

Procedure:

Base Component – Disperse the **Rayox R-88, IAF X-2280, M-P-A** (xylene), and soya lecithin in suitable portion of **Vanoxy 201-X-75 and** solvents. Let down with remaining **Vanoxy 201-X-75**, **Beetle 216-8**, and remaining solvents.

Curing Agent Component – Charge **Van-Amid 300-KX-60** to a suitable container and under constant agitation add **DMP-30** and solvents. Mix thoroughly.

Package base component and curing agent component separately. Mix just prior to use. Allow one hour induction time before applying.

Total NV by Wt. (%)	60.2
P/B Ratio by Wt.	45.3:54.7
lb/gal	10.19
Stormer Visc. (KU)	68

No. 2

(Two-Package, White, Air-Dry)

		lb	gal
A	**Araldite 7071**	243.6	24.60
	RA-50	198.9	5.83
	Xylene	133.2	18.39
	Cellosolve	51.2	6.61

	MIBK	51.2	7.67
	SR-82	1.6	0.18
	Bentone 27	2.4	0.17
B	**Polyamide 815 X-70**	187.9	24.09
	Solvesso 150	92.5	12.46

Total NV (%)	60.0
P/B Ratio	35 : 65
lb/gal at 25 C	9.6

Epoxy-Polyester Enamel

Formula No. 1

(Gloss, Green)

	lb	gal
Part A – Pigment Dispersion		
Aroflint 606	212.5	25.23
M-P-A-60	10.0	1.47
Modaflow (25% in xylol)	4.5	0.61
R-902	162.0	4.99
A-4436	18.0	1.23
Solvent	116.7	16.51
Part B – Clear Activator		
Aroflint 252 MA1-60	433.0	46.34
Solvent	26.0	3.70
Enamel Formula:		
Part A – Pigment Dispersion	523.2	50.04
Part B – Clear Activator	459.0	50.04

PVC (%)	12.0
Total Solids by Wt. (%)	66.7
Total Solids by Vol. (%)	52.5
Visc. (KU)	60–65

No. 2

(High Gloss, White)

	lb	gal
Part A – Pigment Dispersion		
Aroflint 606	125.0	14.88
M-P-A (xylene)	10.0	1.37
Modaflow (25% in xylene)	5.0	0.68
Ti Pure R-902	275.0	8.31
Roller mill 2 passes:		
Aroflint 606	127.0	15.11
Hi-Sol 15	71.4	9.65
Part B – Clear Activator		
Aroflint 404-XA7-60	372.7	36.76
Ethylene Glycol Monoethyl Ether Acetate	107.2	13.24
Enamel Formula:		
Part A – Pigment Dispersion	613.4	50.00
Part B – Clear Activator	479.9	50.00

Visc. (KU)	65-70
PVC (%)	15.5
Total Solids by Vol. (%)	68.7
Total Solids by Wt. (%)	59.5

Epoxy-Urea/Formaldehyde Enamel

Formula No. 1

(Chrome Oxide Green)

		lb	gal
A	Chromium Oxide	175.0	4.05
	Barytes	50.0	1.35
	Gold Bond R Silica	100.0	4.52
	X-2280	2.0	0.09
	M-P-A 60 (xylene)	5.0	0.69

Epi-Rez 2036	440.0	49.00
Syn-U-Tex 402	12.8	1.50
Methyl Ethyl Ketone	67.1	10.00
Methyl Isobutyl Ketone	58.8	8.80
B **Epi-Cure KCX-42-801**	133.5	16.00
Xylol	28.7	4.00

Composite Blend:

Part A	910.7	80.00
Part B	162.2	20.00

lb/gal	11.40
Visc. (KU)	67
PVC (%)	20.3
Total NV by Vol. (%)	53.0
Total NV by Wt. (%)	66.8

No. 2

(High Gloss, White)

	lb/100 gal
Unitane OR-600	224.60
Epon 1007-CT-55	250.20

Disperse and add the following:

Epon 1007-CT-55	65.60
Beetle 1053-10	106.40
Catalyst 4040	1.25
Dow Corning No. 1	0.50
Ethylene Glycol Monoethyl Ether Acetate	173.10
Toluene	68.80
Xylene	7.10
Butanol	78.30

Total NV by Wt. (%)	48.5
Total NV by Vol. (%)	31.1
Visc. (s–Ford 4 Cup)	25

Latex White Enamel

Formula No. 1

(Premium Quality, Gloss)

	lb	gal
Charge the following to the tank:		
Tamol 731 (25%)	10.8	1.31
Nopco NDW	2.0	0.27
Propylene Glycol	59.8	6.94
Add the following at low speed:		
Rutile Titanium Dioxide (high gloss)	269.6	7.86

Grind the above in a high-speed mill (Cowles, 3800-4500 fpm for 10-15 min) and let down at lower speed with the following:

		lb	gal
Propylene Glycol		57.8	6.71
Rhoplex AC-490 (46.5%)		555.9	62.71
Super Ad-It	Premix	1.0	0.12
Water	Premix	15.2	1.82
Texanol	Premix	15.7	1.98
Triton GR-7	Premix	2.0	0.23
Nopco NDW		2.9	0.40
Water and/or Hydroxyethyl Cellulose (2.5% sol'n.)*		80.4	9.65

* Adjust initial viscosity to 70 ± 2 KU.

PVC (%)	22.8
Total Solids by Vol. (%)	34.5
Total Solids by Wt. (%)	49.2
Visc.–Initial (KU)	68-72

No. 2

(Medium Gloss)

	lb	gal
Charge the following to the tank:		
Tamol 731 (25%)	10.4	1.14
Nopco NDW	1.9	0.25
Propylene Glycol	63.4	7.35

Add the following at low speed:

Rutile Titanium Dioxide (high gloss)	260.3	7.59

Grind the above in a high-speed mill (Cowles, 3800-4500 fpm for 10-15 min) and let down at lower speed with the following:

Propylene Glycol		88.4	10.25
Rhoplex AC-490 (46.5%)		398.5	44.95
Nopco NDW		1.9	0.25
Super Ad-It	Premix	1.0	0.12
Water	Premix	92.4	11.08
Butyl **Cellosolve**	Premix	18.5	2.37
Triton GR-7		1.9	0.22
Hydroxyethyl Cellulose (2.5% sol'n.)		120.2	14.43

PVC (%)	28.5
Total Solids by Vol. (%)	26.7
Total Solids by Wt. (%)	42.1
Visc.—Initial (KU)	73

No. 3

(Semigloss)

	lb/100 gal
Water	125.0
Ad-A-Blend PD-919	68.5
Defoamer L-475	2.5
Dowicil 100	1.0
R-900	275.0

Mix

Cellosize QP4400H	4.5

Disperse:

Water	200.8
PD-109-P Latex	400.0
Defoamer L-475	2.0

Procedure:

Formulated for a high-speed disperser. Add in order while stirring at a moderate speed. Mix 3-5 min. Disperse at 4000-4500 fpm for about 20 min. Reduce speed, add last 3 items and mix thoroughly.

lb/gal	10.8
Total NV (%)	47.4
PVC (%)	25
Visc. (KU)	7

Latex Eggshell Enamel

	lb	gal
Charge the following to the tank:		
Tamol 731 (25%)	13.0	1.41
Nopco NDW	2.0	0.27
Propylene Glycol	67.0	7.77
Water	50.0	6.00
Butyl **Cellosolve**	22.0	2.93
Add the following at low speed:		
Rutile Titanium Dioxide	275.0	8.02
Clay	100.0	4.66
Grind the above in a high-speed mill (Cowles, 3800-4500 fpm for 10-15 min) and let down at lower speed with the following:		
Rhoplex AC-490 (46.5%)	473.3	53.77
Nopco NDW	2.0	0.27
Super Ad-It	0.5	0.06
Triton GR-7	2.0	0.23
Water and/or Hydroxyethyl Cellulose (2.5% **QP-4400** sol'n.)*	121.8	14.61

* Adjust initial viscosity to 75 ± 3 KU.

PVC (%)	35.9
Total Solids by Vol. (%)	35.3
Total Solids by Wt. (%)	52.7
Visc.—Initial (KU)	72-78

Polyester White Enamel

(Detergent-Resistant, High Gloss)

Unitane OR-600	295.1
Cyplex 1473-5 (65%)	151.6
Xylol	12.2

Disperse on 3-roll mill and add:

Cyplex 1473-5 (65%)	228.0
Epoxy Resin (low mw, 60% in MIBK)	54.8
Cymel 300	49.3
SR-82	7.0
Dow Corning Paint Additive No. 1	3.6
Xylol	12.4
Butanol	59.1
Methyl Isobutyl Ketone	101.8
Catalyst 1010	3.3
Cellosolve Acetate	20.4
Isophorone	20.6
Pine Oil	20.8

Visc. (s–No. 4 Ford Cup)	15
lb/gal	10.4
Total NV by Wt. (%)	60.8
Total NV by Vol. (%)	40.6
P/B Ratio	90:100

Triamino Ester White Enamel

	lb	gal
Ti Pure R-900	300.0	8.58
Nytal 300	200.0	8.40
Pentachlorophenol	4.0	0.50
Chemacoil TA303 (80%)	306.0	40.00
Thixcin 25C	6.0	0.90
Methyl Ethyl Ketoxime	2.0	0.20
Mineral Spirits	46.0	7.00

Grind 6 N.S. and add:

Chemacoil TA303 (80%)	126.0	16.44
Cobalt Naphthenate (6%)	0.5	0.06
Lead Naphthenate (25%)	5.0	0.50
Mineral Spirits	126.0	19.42

lb/gal	11.0
PVC (%)	28.3
Total Solids by Wt. (%)	75.4
Visc. (KU)	79

Acrylic White Spray Enamel

Formula No. 1

	lb	gal
Roller mill grind:		
Rutile Titanium Dioxide	153.2	4.39
Acryloid B-67 (45%)	90.8	12.62
VM&P Naphtha	11.2	1.79
Mix with		
Acryloid B-67 (45%)	419.9	58.30
VM&P Naphtha	52.5	8.36
Mineral Thinner	94.5	14.54

lb/gal	8.2
Total Solids (%)	46.6
Visc. (s–No. 4 Ford Cup)	77

No. 2

(Gloss, For Electrostatic Application)

Cowles grind:	
Acrysol WS-68 (38%)	12.4
Water	26.5
DMAE (10% in water)	3.3
Tamol 731 (25%)	2.7
Triton X-114	2.0
n-Butanol	4.1

Mix the above ingredients, then add slowly:

RCL-9	134.5

Mix the grind for 20 min, then let down under good agitation as follows:
Let down:

Water	Premix	108.1
Butyl **Cellosolve**	Premix	4.7
n-Butanol	Premix	6.3
Acrysol WS-68 (38%)		414.0
L-5310 (30% in butyl **Cellosolve)**		1.4
Cymel 303		40.4
Dipropylene Glycol		74.7
DMAE (10% in water)		131.8
PTSA (10%)		4.0
Total Solids (%)	35.0	
P/B Ratio	40:60	
pH	8.4	
Visc. (s–No. 4 Ford Cup)	43	

No. 3

(High Solids)

	lb	gal
Rutile Titanium Dioxide	264.6	7.9
XC-4012	264.2	30.4
Dimethylamino Ethanol	4.6	0.6
Demineralized Water	59.8	7.2
Cymel 303	66.3	6.6
BYK-P-104S	3.3	0.4

Disperse above on sand mill. Add the following to mill base using low speed agitation:

XC-4012	279.8	32.2

Premix the following and add:

Catalyst 4040	1.7	0.2
Dimethylamino Ethanol	4.6	0.6

Finally add the following:

Demineralized Water	115.4	13.9

Acrylic-Modified Alkyd Spray Enamel

	lb	gal
Roller mill grind:		
Rutile Titanium Dioxide	277.0	7.95
Zinc Oxide	14.6	0.29
Duraplex D-65A (70%)	194.4	24.63
Mix with:		
Duraplex D-65A (70%)	118.1	14.96
Acryloid B-67 (45%)	162.0	22.53
VM&P Naphtha	20.3	3.21
Mineral Thinner	170.1	26.20
Cobalt Naphthenate (6% metal)	1.8	0.23

lb/gal	9.6
Total Solids (%)	60.8
Visc. (s–No. 4 Ford Cup)*	73

* Enamel should be made to an approximate visc. of 110 s, then allowed to equilibrate for about 3 days and adjusted to 73 s.

Alkyd Aerosol Enamel

Formula No. 1

(White)

	lb	gal
Ti Pure R-960	140.0	4.1
Lankyd 1655	27.5	3.4
Lead Naphthenate (24%)	3.0	0.3
Butyl **Cellosolve**	8.5	1.2
Xylol	33.0	4.6
M-P-A 60 (xylene)	4.0	0.5

Pebble mill to 7-N.S. Bring temperature to 110 F minimum. Keep below 160 F to best incorporate **M-P-A 60** (xylene).

Lankyd 1655	295.0	36.0
Lead Naphthenate (24%)	1.0	0.1
Cobalt Naphthenate (6%)	1.6	0.2

Manganese Naphthenate (6%)	1.6	0.2
Aska	0.8	0.1
Toluene	250.0	34.5
Methylene Chloride	100.0	9.1

PVC (%)	19.1
Total Solids (%)	35.1
Total Vehicle Solids (%)	22.4
lb/gal	9.2
Visc. (s–No. 4 Ford Cup)	14

No. 2

(Fast Dry, Gloss)

	lb	gal
Varcopol 474-60 VM	138.0	18.0
Toluol	87.0	12.0
M-P-A 60	3.0	0.4
Titanium Dioxide	300.0	8.6

Sand grind:

Xylol	65.0	9.0
Varcopol 474-60 VM	306.0	40.0
Toluol	116.0	16.0
Cobalt Naphthenate (6%)	1.8	0.2
Exkin #2	1.0	0.1
DC-200 (1% in xylol)	3.4	0.4

PVC (%)	21.5
Total NV by Wt. (%)	55.7
lb/gal	9.77
Visc. (s–No. 4 Ford Cup)	40–45

Polyester White Spray Enamel

Unitane OR-600	238.0
Cyplex 1473-5 (65%)	122.1
Xylol	9.9

Disperse on 3-roll mill and add:

Cyplex 1473-5 (65%)	162.3
Cellulose Acetate-Butyrate (½ in., 20% solids sol'n. in methyl isobutyl ketone)	198.1
Cymel	39.7
Catalyst 4040	4.6
Methyl Isobutyl Ketone	27.3
Xylol	122.6
Butanol	42.8

Visc. (s–No. 4 Ford Cup)	60
lb/gal	9.7
Total NV by Wt. (%)	52
Total NV by Vol. (%)	27.4
P/B Ratio	90:100
PVC (%)	21.1

Application:

Spray on metal.

Chapter VI

INTERIOR PAINTS

Flat Wall Paints

The important characteristics of a good flat wall paint are: high hiding power; a uniform appearance, resistance to burnishing, staining, scrubbing, chipping, and yellowing; good pigment binding properties; hard and smooth film; and moderate cost.

Acrylic Flat Wall Paint

Formula No. 1

(Thixotropic)

		lb	gal
Charge the following to the tank:			
Water		235.0	28.20
Tamol 731 (25%)		10.0	1.09
Hydroxyethyl Cellulose (2.5% sol'n.)		150.0	18.00
Nopco NXZ		1.0	0.14
Texanol	Premix	15.0	1.89
Ethylene Glycol	Premix	30.0	3.22
Super Ad-It	Premix	0.5	0.06
Add while mixing at low speed, 1900 fpm:			
Ti Pure R-901		210.0	6.30
Imsil A-25		85.0	3.78
Icecap K		100.0	4.62
Attagel 20		30.0	1.52

Grind the above on a high-speed disperser for 15 min at 4500 fpm; then let down the following at low speed, 1900 fpm:

Nopco NXZ	2.0	0.28
Rhoplex AC-388 (50%)	245.0	27.81
Hydroxyethyl Cellulose (2.5% sol'n.)	123.0	14.75

PVC (%)	54.0
Total Solids (%)	46.5
Visc.–Initial (KU)	≈86

No. 2

(High Quality)

	lb	gal
Charge the following to the tank:		
Water	75.0	9.00
KTPP	1.5	–
Tamol 850 (30%)	6.7	0.68
Hydroxyethyl Cellulose (3½%–**QP-300**)	140.0	16.80
Ethylene Glycol	15.0	1.61
Dalpad A	12.5	1.36
Nopco NXZ	2.0	0.27
Add the following at low speed:		
Ti Pure R-911	300.0	8.91
Optiwhite	75.0	4.09
Min-U-Sil	70.0	3.17
Grind for 20 min at high speed on a Cowles then add the following at low speed:		
Rhoplex AC-388 (50%)	418.5	47.71
Super Ad-It	1.0	0.12
Nopco NXZ	1.0	0.14
Water and/or Hydroxyethyl Cellulose 3½% (**QP-300**)*	108.0	12.96

* Adjust initial viscosity to 85 ± 3 KU.
Adjust pH to 9.5 with 28% NH_4OH.

PVC (%)	43.0
Total Solids by Vol. (%)	34.8
Total Solids by Wt. (%)	54.0
Visc.–Initial (KU)	85 ± 3

No. 3

(High Quality)

	lb	gal
Charge the following to the tank:		
Water	111.2	13.34
Tamol 731 (25%)	10.0	1.09
Triton CF-10	2.0	0.22
Hydroxyethyl Cellulose (2.5% sol'n.)	50.0	6.05
Colloid 600	1.0	0.13
Hexylene Glycol	30.0	3.93
Ethylene Glycol	15.0	1.61
Preservative (10% mercury)	1.0	0.12
Add while mixing at low speed, 1900 fpm:		
Rutile Titanium Dioxide (nonchalking)	180.0	5.15
Icecap K	130.0	6.01
Zeolex 80	27.0	1.54
1160 Silica	52.5	2.39

Grind the above on a high-speed disperser for 15 min at 4500 fpm; then let down the following at low speed, 1900 fpm:

	lb	gal
Water	75.0	9.00
Colloid 600	2.0	0.26
Rhoplex AC-388 (50%)	211.5	24.11
Hydroxyethyl Cellulose (2.5% sol'n.)	207.4	24.89

Adjust pH to 9.5 with ammonium hydroxide (28%).

PVC (%)	57
Total Solids (%)	44.8
Visc.–Initial (KU)	≈ 87

No. 4

Water	321.0
KTPP	2.5
Tamol 731 (25%)	5.0
Foamaster VL	2.0
Dowicil 75	1.3
Ethylene Glycol	19.0
Rutile Titanium Dioxide	170.0
Icecap K	150.0
Duramite	75.0
Silica 1160	92.0
Attagel 50	6.0
Natrosol 250 HR	5.0

Grind the above at high speed for 15 min, then let down at a slower speed as follows:

Foamaster VL		6.0
Rhoplex AC-64 (60.5%)		210.2
Texanol		8.8
Water	Premix	53.6
Natrosol 250 HR (2½%)	Premix	55.0

PVC (%)	59.5
Total Solids by Vol. (%)	33.4
pH—Initial	9.5
Visc.—Initial (KU)	98

No. 5

(Eggshell)

	lb	gal
KTPP	2.0	0.09
Tamol 850 (30%)	6.0	0.66
Triton CF-10	2.0	0.22
Colloid 691	1.0	0.14
Propylene Glycol	61.0	7.08

Water		60.0	7.20
Zopaque RCL-9		300.0	8.76
Imsil A-15		100.0	4.53

Grind on Cowles at high speed (3900 fpm) for 20 min.

Rhoplex AC-64 (60.5%)		400.0	44.60
Propylene Glycol	Premix	6.0	0.70
Texanol		12.1	1.53
Dowicil 75	Premix	1.5	0.11
Water		43.0	5.16
Colloid 691		3.0	0.54
Acrysol G-110 (22%)	Premix	10.0	1.13
Water		25.0	3.00
Natrosol 250 GR (4½% sol'n.) and/or Water		121.3	14.55

PVC (%)	33.9
Total Solids by Vol. (%)	39.2
Visc.—Initial (KU)	80–85
pH	9.3

No. 6

(Eggshell)

KTPP	2.0
Tamol 850 (30%)	6.0
Triton CF-10	2.0
Foamaster L	2.0
Propylene Glycol	61.0
Water	80.0
Unitane OR-600	300.0
Imsil A-15	100.0

Grind the above for 15–20 min at high speed, then let down as follows:

Rhoplex AC-64 (60.5%)		400.0
Propylene Glycol		6.0
Foamaster L	Premix	6.0
Texanol		12.1
Dowicil 75	Premix	1.2
Water		25.0

Collacral VL	Premix	30.0
Water		30.0
Acrysol ASE-60	Premix	12.0
Water		79.7
Ammonia (28%)		3.0

PVC (%)	34
Total Solids by Vol. (%)	39.2
pH	9.0-9.4
Visc.–Initial (KU)	78-82

Alkyd Flat Wall Paint

Formula No. 1

	lb	gal
Dispersion:		
Ti Pure R-931	200.0	6.00
Camel-Wite	460.0	20.38
Celite 281	50.0	2.60
Soya Lecithin	7.0	0.81
Bentone 38 (10% gel)*	45.0	6.34
Varkyd 558	200.0	27.40
Reduction:		
Varkyd 558	222.0	30.43
Lead Naphthenate (24%)	1.8	0.19
Cobalt Naphthenate (6%)	0.8	0.10
Antiskinning Agent	1.0	0.11
Mineral Spirits (heavy)	37.0	5.64

PVC (%)	65.0
Visc.–Overnight (KU)	85-90

Disperse in high-speed dispenser

* **Bentone 38**	32
Mineral Spirits (heavy)	272
Proprietary Alcohol	16

No. 2

(High Hiding)

	lb	gal
Titanox RCHT	650.0	23.95
Celite 281	100.0	5.20
Bentone 38	3.0	–
Low Solids Alkyd	317.5	45.00
TT-R-266a Type I Alkyd	24.0	3.00
Mineral Spirits (low odor)	144.0	22.00
Cobalt Naphthenate (6%)	4.0	0.50
Zirco Drier (6%)	4.0	0.50
Antiskinning Agent	1.0	0.13
Pigment (%)	60.3	
Vehicle (%)	39.7	
Vehicle NV (%)	22.6	

No. 3

Rutile Titanium Calcium	400.0
Calcium Carbonate	350.0
Wallkyd 1956-35	352.0
Mineral Spirits (odorless)	126.0
Lead Naphthenate (24%)	2.0
Cobalt Naphthenate (6%)	0.5
Antiskinning Agent	1.5

No. 4

(Low Odor)

Rutile Titanium Dioxide	2.86
Titanium Calcium Pigment (30% TiO_2)	11.10
Calcium Carbonate (ground limestone)	11.10
Alkyd Resin **3113-P-30** (30% solids)	42.68
Alkyd Resin **3113-P-30**	16.42
Lead Naphthenate (24%)	0.20

Cobalt Naphthenate (6%)	0.13
Antiskinning Agent	0.13
Mineral Spirits (low odor)	15.38

No. 5

(Odorless)

	lb	gal
Ti Pure R-300	275.0	7.9
Atomite	125.0	5.5
Snowflake	250.0	11.1
Thixatrol ST	4.0	0.5
Mineral Spirits (odorless)	90.0	14.0
Varkyd 1533-50 O.S.	192.0	25.0

Mix and heat to 135 F roller mill

	lb	gal
Varkyd 1533-50 O.S.	100.0	13.0
Mineral Spirits (odorless)	142.0	22.0
Cobalt Octoate (6%)	1.5	0.2
Lead Octoate (24%)	3.0	0.3
Exkin #2	1.0	0.1

PVC (%)	60.8
Vehicle NV by Wt. (%)	28.4
Total NV by Wt. (%)	67.7
lb/gal	11.9
Visc. (KU)	87

Latex Flat Wall Paint

Formula No. 1

	lb	gal
Water	50.0	6.00
Daxad 30	7.2	0.75
Texanol	4.8	0.60
Ethylene Glycol	23.3	2.50
Super Ad-It	0.3	0.03

Nopco NXZ	1.9	0.25
Triton X-100	3.0	0.34
Natrosol 250 HR (2% sol'n.)	200.0	24.00
Rutile Titanium Dioxide	200.0	5.71
Gold Bond "R" Silica	100.0	4.55
Micro Cel T-70	50.0	2.50
Albacar 41-8	200.0	8.88

Grind above in Cowles dissolver; then premix following and blend with **Poly-Tex 668** and dibutyl phthalate before mixing with grind. Use gentle agitation.

Poly-Tex	186.0	20.00
Dibutyl Phthalate	3.2	0.37
Water	144.0	17.27
Nopco NXZ	1.9	0.25

Use following as needed for viscosity adjustment:

Water (or protective colloid sol'n.)	50.0	6.00

Visc.–Initial (KU)	89
lb/gal	12.26
PVC (%)	65
Total Vehicle Solids (%)	17.9
Total Solids (%)	54.8

No. 2

(White)

	lb	gal
Water	50.0	6.00
Tamol 731	6.5	0.75
Igepal CO-630	4.4	0.50
Ethylene Glycol	18.6	2.00
Colloid 581 B	3.4	0.50
Texanol	5.3	0.67
Super Ad-It	0.3	0.03
Rutile Titanium Dioxide (NC)	150.0	4.28
Calcined Clay	100.0	4.67

Duramite	200.0	8.87
Celite 281	50.0	2.61
Cellosize QP-4400 (2½% sol'n.)	200.0	24.00

Grind above in 3 roll mill (3 pcs); then add following under gentle agitation:

Poly-Tex 660	182.0	20.00
Water	209.0	25.12
Visc.—Overnight (KU)	75	
lb/gal	11.80	
PVC (%)	67	
Volume NV (%)	16	
Total NV (%)	51	

No. 3

	lb	gal
Water	125.0	15.0
KTPP	1.5	0.1
Igepal CO-530	3.0	0.4
Carbitol Acetate	12.0	1.4
Ethylene Glycol	15.0	1.6
Troy San PMA-30	0.3	—
Titanox RA-50	200.0	5.7
Al-Sil-Ate W	50.0	2.3
Camel Carb	200.0	8.8
Celite 281	25.0	1.3
Nopco NDW (defoamer)	1.0	0.1

Disperse

Nopco NDW (defoamer)	1.0	0.1
Natrosol 250 (2% sol'n.)	220.0	26.5
40-124 **Wallpol**	220.0	24.0
Water	125.0	15.0
Total NV	50.6	
pH	7.6	
Visc. (KU)	84	
PVC (%)	60	
lb/gal	11.7	

No. 4

	lb/100 gal
Water	250
Tamol 731 (25%)	8
Nopco NDW	2
Dowicil 100	1
Triton X100	5
Ti Pure R-901	250
ASP 400	100
Snowflake White	125
Ethylene Glycol	25
Dalpad A	10
Everflex E (55%)	250
Water	142
XD-7845.01	8

	No. 5 (White Emulsion)		No. 6 (White Emulsion)	
	lb	gal	lb	gal
Ti Pure R-901	200.0	6.0	200.0	6.0
Drikalite	450.0	20.0	450.0	20.0
Daxad 30	5.0	0.5	5.0	0.5
Colloid 606	2.0	0.3	2.0	0.3
Ethylene Glycol	28.0	3.0	28.0	3.0
Hexylene Glycol	7.0	1.0	7.0	1.0
Troysan PMA-30	0.3	–	0.3	–
Igepal CO-610	2.0	0.2	2.0	0.2
Natrosol 250 HR (3%)	75.0	9.0	75.0	9.0
Water	192.0	23.0	216.0	26.0
Everflex G (52 NV)	329.0	37.0	–	–
Flexbond 840 (56 NV)	–	–	310.0	34.0
PVC (%)	59.7		59.6	
Visc. (KU)	80–85		80–85	
Paint NV by Vol. (%)	45.9		46.0	
Paint NV by Wt. (%)	63.7		64.0	
Vehicle NV by Wt. (%)	27.4		27.6	

No. 7

(Vinyl Chloride Modified)

	lb	gal
Titanox RANC	175.7	5.0
ASP 400	165.7	7.7
Celite 281	88.5	4.5
Atomite	80.9	3.6
KTPP	2.0	–
Igepal CO-630	2.0	0.2
Water	190.7	22.8
Cellosize WP-4400 (2½% sol'n.)	219.7	26.2
Texanol	3.0	0.4
Ethylene Glycol	18.5	2.0
Geon 450X20	160.7	17.9
Colloids 60	2.0	0.2
2-Amino-2-Methyl-1-Propanol	2.0	0.2
Water or **Cellosize** Sol'n.	102.9	12.3

PVC (%)	70.41
Total NV (%)	49.95
NV by Vol. (%)	28.73
lb/gal	11.78
Visc. (KU)	74.00
pH	9.10

No. 8

(White)

	lb	gal
Dispersion		
Water	400.0	48.02
Natrosol 250 MR	5.0	0.45
PMA-18	3.0	0.30
KTPP	0.5	0.23
Daxad 30 (25%)	4.0	0.41

Igepal CO-630	2.0	0.23
Ethylene Glycol	25.0	2.70
Texanol	5.0	0.63
Unitane OR-560	150.0	4.39
Micro-Cel T-70	20.0	1.06
Opti White	50.0	2.72
ASP 400	150.0	6.98
Atomite	125.0	5.54

Reduction:

Everflex E (55%)	165.0	18.33
Water	75.0	9.00

PVC (%)	68.9
Total Solids (%)	50.4
Total Solids by Vol. (%)	30.6
Visc. (KU)	85 ± 5

No. 9

	lb/100 gal
Water	163
Tamol 731 (25%)	8
Igepal CO-630	5
Ti Pure R-901	250
ASP 170	100
Atomite	100
Drew 913 BL	1
Triton GR-7	3
Dow Latex 308 (55%)	245
Propylene Glycol	10
Ethylene Glycol	15
Dowicil 100	2
Drew 913 BL	2
Water	227
XD-7846.01	

No. 10

		lb	gal
Water		230.0	27.64
Daxad (25%)		8.0	0.82
R&R 551		5.0	0.58
Dowicil 100		2.0	0.17
Ti Pure R-911		300.0	8.72
Satintone #1		100.0	4.62
Gold Bond "R"		50.0	2.27
Celite 281		30.0	1.56
Snowflake		25.0	1.11
Antifoamer (50%)		2.0	0.25
Propylene Glycol	Premix	75.0	8.68
Methocel 90 HG (4000 cps)	Premix	4.2	0.42

Let down:

	lb	gal
Water	79.1	9.50
Elvanol 52-22 (15%)	40.0	4.79
Dow Latex 308 (55%)	233.0	27.48
Defoamer 913 BL	4.0	0.51
Mineral Spirits	6.0	0.90

PVC (%)	55.0
Total Solids by Wt. (%)	55.0
Total Solids by Vol. (%)	33.0

No. 11

(Premium-Quality)

Charge the following to the tank:

	lb	gal
Water	66.8	8.02
KTPP	1.4	–
Tamol 850 (30%)	6.2	0.63
Hydroxyethyl Cellulose–3½% (**QP-300** sol'n.)	111.4	13.37
Ethylene Glycol	27.8	2.99
Dalpad A	18.6	2.02
Nopco NXZ	1.9	0.25

Add the following at low speed:

Rutile Titanium Dioxide	278.5	8.27
Clay	69.6	3.80
Min-U-Sil 30	65.0	2.94

Grind the above in a high-speed mill (Cowles, 3800–4500 fpm for 10–15 min) and let down at lower speed with the following:

Rhoplex AC-490 (46.5%)	417.8	47.22
Nopco NXZ	0.9	0.13
Super Ad-It	0.9	0.11
Water and/or Hydroxyethyl Cellulose–3½% (**QP-300** sol'n.)*	85.4	10.25

* Adjust initial visc. to 85 ± 3 KU.
Adjust pH to 9.5 with 28% NH_4OH.

PVC (%)	42.9
Total Solids by Vol. (%)	35.0
Total Solids by Wt. (%)	53.3
Visc.–Initial (KU)	82-88
pH–Initial	9.5

No. 12

(High Quality)

Charge the following to the tank:

	lb	gal
Water	72.0	8.64
KTPP	1.5	–
Tamol 850 (30%)	6.7	0.68
Hydroxyethyl Cellulose–3½% (**QP-300**)	120.0	14.40
Ethylene Glycol	30.0	3.22
Dalpad A	20.0	2.18
Nopco NXZ	2.0	0.27

Add the following at low speed:

Ti Pure R-911	303.0	8.91
Optiwhite	75.0	4.09
Min-U-Sil 30	70.0	3.17

Grind for 20 min at high speed on a Cowles then add the following at low speed:

Rhoplex AC-490 (46.5%)	450.0	50.85
Super Ad-It	1.0	0.12
Nopco NXZ	1.0	0.14
Water and/or Hydroxyethyl Cellulose—3½% (**QP-300**)*	92.0	11.04

* Adjust initial visc. to 85 ± 3 KU.
Adjust pH to 9.5 with 28% NH_4OH.

PVC (%)	42.6
Total Solids by Vol. (%)	35.2
Total Solids by Wt. (%)	53.3
Visc.—Initial (KU)	85 ± 3

No. 13

(High Quality)

Charge the following to the tank:

	lb	gal
Water	62.0	7.44
KTPP	1.5	—
Tamol 850 (30%)	6.7	0.68
Hydroxyethyl Cellulose—3½% (**QP-300**)	130.0	15.60
Ethylene Glycol	15.0	1.61
Hexylene Glycol	35.0	4.58
Nopco NXZ	2.0	0.27

Add the following at low speed:

Ti Pure R-911	300.0	8.91
Optiwhite	75.0	4.09
Min-U-Sil 30	70.0	3.17

Grind for 20 min at high speed on a Cowles then add the following at low speed:

Rhoplex AC-22 (44.5%)	470.0	53.11
Super Ad-It	1.0	0.12
Nopco NXZ	1.0	0.14
Water and/or Hydroxyethyl Cellulose–3½% **(QP-300)***	57.0	6.84

---

* Adjust initial visc. to 85 ± 3 KU.
Adjust pH to 9.5 with 28% ammonia.

PVC (%)	42.7
Total Solids by Vol. (%)	35.3
Total Solids by Wt. (%)	53.8
Visc.–Initial (KU)	85 ± 3

No. 14

(High-Film Build, Washable)

	lb	gal
Water	100.0	10.80
Tamol 731 (25%)	14.0	1.53
Triton CF-10	2.0	0.22
Hydroxyethyl Cellulose–2.5% **(QP-4400)**	60.0	7.27
Nopco NXZ	2.0	0.28

Add while mixing:

Ethylene Glycol	10.0	1.08
Hexylene Glycol	30.0	3.93
Super Ad-It	1.0	0.12

Add:

Ti Pure R-901	300.0	9.03
Optiwhite	86.0	4.68
1160 Silica	57.0	2.59

Grind on high-speed dispersion equipment for 15 min, then add at low speed:

Nopco NXZ	2.0	0.28
Rhoplex AC-22	397.0	44.80
Hydroxyethyl Cellulose–2.5% **(QP-4400)**	81.0	9.81
Water	25.0	4.20

Tone with California Ink 8800 line:
0.44 lb Colorant B/100 gal and 0.66 lb Colorant L/100 gal
Adjust pH to 9.5 with ammonium hydroxide (28%).

PVC (%)	45
Total Solids (%)	53.5
Visc.–Initial (KU)	87-93

No. 15

(Moderate-Film Build, Washable)

	lb	gal
Water	100.0	12.00
Tamol 731 (25%)	14.0	1.53
Triton CF-10	2.0	0.22
Kelzan/Keltex P (6/1, 4% solids)	60.0	7.26
Nopco NXZ	2.0	0.27

Add while mixing:

Ethylene Glycol	10.0	1.08
Hexylene Glycol	30.0	3.93
Super Ad-It	1.0	0.12

Add:

Ti Pure R-901	300.0	9.03
Optiwhite	86.0	4.69
1160 Silica	57.0	2.58

Grind on high-speed dispersion equipment for 15 min, then add at low speed:

Nopco NXZ	2.0	0.27
Rhoplex AC-22	397.0	44.86
Kelzan/Keltex P (6/1, 4% solids)	4.0	0.48
Water	102.0	12.24

Tone with California Ink 8800 line:
0.44 lb Colorant B/100 gal and 0.66 lb Colorant L/100 gal
Adjust pH to 9.5 with ammonium hydroxide (28%).

PVC (%)	45
Total Solids (%)	53.5
Visc.–Initial (KU)	75-81

PV Acetate Interior Flat Paint

(Medium-Quality)

Charge the following to the tank in order of addition:

		lb/100 gal
Zopaque RCS-3 (67% solids)		224
Kaopaque 20 (65.4% solids)		156
Camelwhite (71.7% solids)		306
Ethylene Glycol		25
Nopco NDW		
Flexbond-315 (55% solids)		281
Water	Premix before adding	23
Troysan CMP Acetate		1
Carbitol Acetate		10
Hydroxyethyl Cellulose–2½% sol'n. **(QP-4400)**		160

PVC (%)	54
lb/gal	11.9
Total NV by Wt. (%)	52.8
Visc.–Sheared (KU)	92

PV Acrylic Interior Flat Wall Paint

Formula No. 1

(Low-Cost)

	lb/100 gal
Water	160.0
Ethylene Glycol	25.0
CMP Acetate	2.5

KTPP	0.5
Tamol 731	8.0
Igepal CO-610	4.0
Cellosize QP-4400 (2½% sol'n.)	50.0
Antifoamer	2.0
Ti Pure TiO_2	75.0
Altowhite LL	175.0
Kaopaque 20	75.0
Duramite	100.0

Grind for 15 min with a Cowles-type mixer at 3800-4500 fpm and let down at slower speed.

Antifoamer	1.0
Carbitol Acetate	6.0
Cellosize QP-4400 (2½% sol'n.)	250.0
Polyvinyl Acrylic Emulsion (55% solids)	136.0
Water	53.0

PVC (%)	70
Total Solids (%)	45
lb/gal	11.2
Visc. (KU)	90-95

No. 2

(Low-Cost)

	lb/100 gal
Water	260.0
Ethylene Glycol	20.0
CMP Acetate	2.5
KTPP	1.0
Tamol 731	10.0
Igepal CO-610	4.0
Cellosize QP-4400 (2½% sol'n.)	50.0
Antifoamer	2.0
Ti Pure TiO_2	50.0
Altowhite LL	175.0
Kaopaque 30	150.0
Duramite	150.0

Grind for 15 min with a Cowles-type mixer at 3800-4500 fpm and let down at slower speed.

Antifoamer	1.0
Texanol	10.0
Cellosize QP-4400 (2½% sol'n.)	155.0
Polyvinyl Acrylic Emulsion (55% solids)	140.0
PVC (%)	75
Total Solids (%)	52
lb/gal	11.8
Visc. (KU)	85-90

Styrene-Acrylic Interior Flat Paint

Formula	No. 1	No. 2 *(High Scrub Resistant)*	No. 3 *(High Opacity)*
Runa RE376	200.0	–	–
Tioxide R-XL	–	200.0	175.0
Microcal ET	–	–	20.0
Micromica W1	56.0	–	–
Whiting **BWF42**	159.0	–	296.0
Microdol No. 1	–	235.0	–
Vinapol 1640 (diluted 1:6 with water)	40.0	–	15.0
Calgon S (5% aq. sol'n.)	–	20.0	15.0
Celacose CP10 (2% aq. sol'n.)	200.0	–	–
Celacol HPM 15000DS	–	250.0	200.0
Potassium Hydroxide (50% aq. sol'n.)	2.0	–	–
Ammonia (sp. gr. 0.880-0.885)	–	2.0	4.0
Water	130.6	122.6	136.6
Foamaster NXZ	2.0	2.0	–
Butyl Dioxitol	20.0	20.0	20.0
Acticide MPM	0.4	0.4	0.4
Vinacryl 7172	190.0	128.0	118.0
Empilan 2020 (10% aq. sol'n.)	–	20.0	–
PVC (%)	60	70	76
Total NV (%)	51	50	55

	No. 4 *(Long Life)*	No. 5 *(General-Purpose)*
Tioxide R-CR2	200	250.0
Microcal ET	–	10.0
Micromica W1	50	–
Microdol No. 1	127	–
D. Powder (China clay)	–	145.0
Celacose CP10 (2% aq. sol'n.)	93	160.0
Vinapol 1640 (diluted 1:6 with water)	40	30.0
Calgon S (5% aq. sol'n.)	–	–
Potassium Hydroxide (50% aq. sol'n.)	2	–
Propylene Glycol	20	–
Water	–	68.6
Ammonia (sp. gr. 0.880–0.885)	–	2.0
Dalpad A	15	15.0
Foamaster NXZ	2	2.0
Acticide MPM	5	0.4
Vinacryl 7172	446	297.0
Empilan 2020 (10% aq. sol'n.)	–	20.0
PVC (%)	35	50
Total NV (%)	60	55

	No. 6 *(Low Cost Scrubbable)*	No. 7 *(Low Cost Washable)*	No. 8 *(Sand Paint)*
Tioxide RXL	75	70	–
Kronos RN44	–	–	100
Microcal ET	10	20	10
D Powder (China clay)	100	–	75
Microdol No. 1	277	–	–
Whiting **BWF42**	–	358	–
Calgon S (5% aq. sol'n.)	15	15	–
Vinapol 1640 (diluted 1:6 with water)	–	–	25
Celacol HPM15000DS (2% aq. sol'n.)	250	–	–

Bermocoll E451FQ (2% aq. sol'n.)	–	315	–
Bermocoll E320GK (4% aq. sol'n.)	–	–	50
Water	157	160	28
Ammonia (sp. gr. 0.880–0.885)	2	–	2
Butyl Dioxitol	15	15	15
Acticide APA	–	–	3
Foamaster NXZ	1	1	2
Formalin	2	2	–
Vinacryl 7172	76	44	250
Empilan 20 20 (10% aq. sol'n.)	20	–	–
Microdol 40-200	–	–	440
PVC (%)	82.5	90	–
Total NV (%)	50	47	75
P/B Ratio	–	–	5:1

Styrene-Acrylic Ceiling or Wall Paint

(Washable)

Premix:

Bermocoll E320G (5% aq. sol'n.)	70
Water	35
Vinapol 1640	10
Propylene Glycol	10
Proxel CRL	1

Add and disperse:

Omya D5	372
Omya D65	300
Tioxide R-CR2	30

Then add:

Butyl Dioxitol	20
Vinacryl 7172	140
Paraplex WP1	10
Foamaster NXZ	2

Styrene-Acrylic Interior Ceiling Paint

(Wipe-Resistant)

Premix:

Bermocoll E320G (5% aq. sol'n.)	70
Water	59
Vinapol 1640	11
Propylene Glycol	10
Proxel CRL	1

Add and disperse:

Omya D5	382
Omya D65	320
Tioxide R-CR2	30

Then add:

Butyl Dioxitol	20
Vinacryl 7172	80
Paraplex WP1	15
Foamaster NXZ	2

Vinyl Acetate Exterior-Interior Flat Paint

Tioxide R-CR2	250.0
'O' Grade Talc	62.0
Whiting **BWF42**	62.0
Modocoll EK600 (4% aq. sol'n.)	140.0
Calgon S (5% aq. sol'n.)	20.0
Vinapol 1640 (diluted 1:6 with water)	20.0
Potassium Hydroxide (50% aq. sol'n.)	2.0
Butyl Dioxitol Acetate	10.0
Foamaster NXZ	1.0
Acticide MPM	0.4
Water	179.6
Vinamul 6705	253.0

PVC (%)	50
Total NV (%)	50
Visc. (at 25 C–poise)	≈ 5
pH	7.8

Vinyl-Acrylic Interior Flat Wall Paint

(High Cost)

	lb/100 gal
Pigment grind:	
Methocel 65 HG—4000 cps (3% sol'n.)	75.0
Tamol 850	7.0
Igepal CO-610	4.0
Ethylene Glycol	25.0
Dee-Fo 97-2	2.0
Texanol	10.0
Merbac 35	0.5
Ti Pure R-901	250.0
Altowhite LL	75.0
M-60	150.0
Celite 499	25.0
Reduction:	
Dee-Fo 97-2	2.0
Vinyl Acrylic (55% solids)	250.0
Total Water and/or **Methocel 65 HG**—4000 cps (3% sol'n.)	340.0
PVC (%)	55
Total Solids (%)	51

Acrylic White Interior Topcoat

Formula No. 1

(Semigloss)

	lb	gal
Propylene Glycol	80.0	9.28
Tamol 731 (25%)	5.0	0.55
Nopco NDW	2.0	0.26
Rutile Titanium Dioxide	225.0	6.57
Silica (small particle, amorphous)	25.0	1.13

Grind the above in a high-speed mill (Cowles) for 20 min and let down at a slower speed as follows:

Water		75.0	9.00
Nopco NDW		4.0	0.52
Rhoplex AC-25 (46.5%)		500.0	56.50
Propylene Glycol	Premix	20.0	2.32
Texanol		14.0	1.77
Water	Premix	16.6	2.00
Dowicil 75		1.3	–
Water/**Natrosol 250 MR** (2½% sol'n.)		84.1	10.10

PVC (%)	24
Total Solids by Vol. (%)	31.8
Visc.–Initial (KU)	72-75
pH	8.5-9.0

	No. 2 *(High Quality, Unmodified)*		No. 3 *(High Quality, Alkyd Modified)*	
	lb	gal	lb	gal
Hydroxyethyl Cellulose (2.5% sol'n.)	85.0	10.30	85.0	10.30
Water	68.5	8.22	68.5	8.22
Tamol 731 (25%)	15.0	1.63	15.0	1.63
Triton CF-10	2.5	0.28	2.5	0.28
Antifoamer	1.0	0.13	1.0	0.13
Preservative	1.8	0.11	1.8	0.11
Ethylene Glycol	25.0	2.69	25.0	2.69
Rutile Titanium Dioxide (nonchalking)	250.0	7.60	250.0	7.60
Talc	203.7	8.80	203.7	8.80

Grind the above materials in a high-speed mill (Cowles, 3800-4500 fpm for 10-15 min) and let down, at a slower speed, as follows:

Rhoplex AC-388 (50%)		459.8	52.23	390.8	44.39
Long Oil Alkyd*		–	–	30.8	3.69
Antifoamer		1.0	0.13	1.0	0.13
Ammonium Hydroxide (28%)	Premix	2.0	0.27	2.0	0.27
Tributyl Phosphate	Premix	11.5	1.41	9.8	1.21
Propylene Glycol	Premix	35.0	4.07	35.0	4.07

Water and/or Hydroxyethyl Cellulose (2.5%)	17.5	2.13	53.5	6.48
Toner: Cal/Ink **GP8814E** Phthalo Blue	0.05	–	0.05	–
Cal/Ink **GP8807B** Lamp Black	0.05	–	0.05	–

* Drier treated with 0.5% by weight of 6% cobalt, 0.5% of 6% manganese, and 1.4% of 24% lead, prior to incorporation of alkyd.

PVC (%)	40.0	40.0
Total Solids by Vol. (%)	41.0	41.0
Total Solids by Wt. (%)	58.0	57.8
pH	≈ 9.5	≈ 9.5
Visc.–Initial (KU)	≈ 72-75	≈ 72-75

No. 4

(Semigloss)

lb/100 gal

Grind the following in a high-speed mill (Cowles, 3800–4500 fpm for 15 min):

Water	20
Cellosize QP-4400 (2.5% sol'n.)	10
Tamol 731 (25%)	18
Defo-233	2
Propylene Glycol	62
Ti Pure R-900	225
Hydrite-UF	43

Let down at lower speed the following:

Propylene Glycol	105
Rhoplex AC-440 (46.5%)	488
Defo-233	2
Super Ad-It	1
Water	10

Butyl **Cellosolve**	23
Triton GR-7	2
Cellosize QP-440 (2.5% sol'n.)	60

PVC (%)	27
Total Solids (%)	32
Visc. (KU)	80–85

No. 5

(Semigloss)

		lb	gal
Charge the following to the tank:			
Tamol 731 (25%)		10.0	1.07
Nopco NDW		1.0	0.14
Propylene Glycol		63.0	7.30
Add while mixing at low speed, 1900 fpm:			
Rutile Titanium Dioxide (high gloss)		225.0	6.57
Barytes		25.0	0.68

Grind the above on a high-speed disperser for 15 min at 4500 fpm; then let down the following at low speed, 1900 fpm:

		lb	gal
Propylene Glycol		100.0	11.60
Rhoplex AC-388 (50%)		457.0	52.10
Nopco NDW	Premix	2.0	0.27
Preservative (10% mercury)	Premix	1.0	0.12
Water	Premix	60.0	7.20
Triton GR-7	Premix	2.0	0.22
Acrysol G-110	Premix	25.0	2.80
Water	Premix	93.0	11.16

PVC (%)	22.8
Total Solids (%)	45.6
Visc.–Initial (KU)	≈ 64

No. 6

(Semigloss, Premium Quality)

Propylene Glycol		93.3
Tamol SG-1 (35%)		8.8
Deefo 495		2.2
Zopaque RCL-9		263.3
Imsil A-15		38.4

Mill the above in a high-speed mill (Cowles) for 20 min and let down at a slower speed as follows:

Water		15.0
Rhoplex AC-417		510.2
Acrysol RM-3 (30%)		34.0
Deefo 495		2.7
Propylene Glycol		16.5
Texanol	Premix	21.9
Super Ad-It	Premix	1.0
Triton GR-7M	Premix	0.5
Water	Premix	14.8
Acrysol RM-4 (30%)	Premix; Premix	20.0
Water	Premix; Premix	40.0
Ammonium Hydroxide (28%)	Premix	0.4
Water	Premix	11.4

PVC (%)	26.5
Total Solids by Vol. (%)	35.7
Visc.—Initial (KU)	70–73
pH	8.2-8.5

No. 7

(Semigloss, Premium Quality)

Propylene Glycol	92.8
Tamol SG-1 (35%)	8.7
Deefo 495	2.2
Zopaque RCL-9	262.2
Imsil A-15	38.2

Mill the above in a high-speed mill (Cowles) for 20 min and let down at a slower speed as follows:

Water		30.0
Rhoplex AC-417 (48%)		507.9
Acrysol RM-3 (30%)		33.9
Deefo 495		2.7
Propylene Glycol	Premix	16.4
Texanol	Premix	7.0
Butyl **Carbitol**	Premix	23.0
Super Ad-It	Premix	1.0
Triton GR-7M	Premix	0.5
Water	Premix	14.7
Water/**Natrosol 250 MR** (2.5% sol'n.)		50.7

PVC (%)	26.5
Total Solids by Vol. (%)	35.5
Visc.–Initial (KU)	70–73
pH	8.2-8.5

No. 8

Natrosol 250 MHR	Premix	3.0
Ethylene Glycol	Premix	25.0
Water		120.0
Tamol 850 (30%)		10.5
KTPP		1.5
Triton CF-10		2.5
Drew L-475		1.0
Propylene Glycol		34.0
Ti Pure R-960		250.0
AZO-11		50.0
Asbestine 3X		178.9

Grind for 10–15 min on Cowles dissolver at 3800–4500 fpm and let down at slower speed as follows:

Rhoplex AC-64 (60.5%)	380.0
Drew L-475	3.0
Texanol	11.0
Skane M-8	2.0
Ammonium Hydroxide	2.0
Water	116.3
Natrosol 250 MHR (2½%)	20.0

PVC (%)	40.0
Total Solids by Vol. (%)	41.0
Visc.–Initial (KU)	75-80
pH	9.5

No. 9

(Dripless)

	lb	gal
Hydroxyethyl Cellulose (2.5% sol'n.)	50.00	6.06
Water	80.00	9.60
Tamol 731 (25%)	15.00	1.63
Triton CF-10	2.50	0.28
Antifoamer	1.00	0.13
Ethylene Glycol	25.00	2.69
Propylene Glycol	35.00	4.07
Preservative	1.80	0.11
Rutile Titanium Dioxide (nonchalking)	250.00	7.60
Talc	203.70	8.80
Magnesium Aluminum Silicate	5.00	0.25

Grind the above materials in a high-speed mill (Cowles, 3800-4500 fpm for 10–15 min) and let down, at a lower speed, as follows:

		lb	gal
Rhoplex AC-388 (50%)		459.80	52.23
Antifoamer	Premix	1.00	0.13
Tributyl Phosphate	Premix	11.50	1.41
Ammonium Hydroxide (28%)	Premix	2.00	0.27
Water and/or Hydroxyethyl Cellulose (2.5%)		39.20	4.74
Toner: Cal/Ink **GP8814E** Phthalo Blue		0.05	–
Cal/Ink **GP8807B** Lamp Black		0.05	–

PVC (%)	40.4
Total Solids by Vol. (%)	41.2
Total Solids by Wt. (%)	58.2
pH	≈9.4-9.7
Visc.–Initial (KU)	≈76-82

No. 10

(Alkyd-Modified, Dripless)

	lb	gal
Hydroxyethyl Cellulose (2.5% sol'n.)	30.00	3.66
Water	80.00	9.60
Tamol 731 (25%)	15.00	1.63
Triton CF-10	2.50	0.28
Antifoamer	1.00	0.13
Ethylene Glycol	25.00	2.69
Propylene Glycol	35.00	4.07
Preservative	1.80	0.11
Rutile Titanium Dioxide (nonchalking)	250.00	7.60
Talc	203.70	8.80

Grind the above materials in a high-speed mill (Cowles, 3800–4500 fpm for 5 min) and then add:

Magnesium Aluminum Silicate	15.00	0.75
Water	50.00	6.00

Grind for 10 additional min and let down, at a lower speed, as follows:

Rhoplex AC-388 (50%)		390.80	44.39
Long Oil Alkyd*		30.80	3.69
Tributyl Phosphate	Premix	9.80	1.21
Ammonium Hydroxide (28%)	Premix	2.00	0.27
Antifoamer	Premix	1.00	0.13
Water and/or Hydroxyethyl Cellulose (2.5%)		41.40	5.01
Toner: Cal/Ink **GP8814E** Phthalo Blue		0.05	–
Cal/Ink **GP8807B** Lamp Black		0.05	–

* Drier treated with 0.5% by weight of 6% cobalt, 0.5% of 6% manganese, and 1.4% of 24% lead, prior to incorporation of alkyd.

PVC (%)	41.1
Total Solids by Vol. (%)	41.8
Total Solids by Wt. (%)	58.6
pH	≈9.4–9.7
Visc.–Initial (KU)	≈77–83

No. 11

(Alkyd-Modified, Semidripless)

	lb	gal
Hydroxyethyl Cellulose (2.5% sol'n.)	80.00	9.70
Water	100.00	12.00
Tamol 731 (25%)	15.00	1.63
Triton CF-10	2.50	0.28
Antifoamer	1.00	0.13
Ethylene Glycol	25.00	2.69
Preservative	1.80	0.11
Rutile Titanium Dioxide (nonchalking)	250.00	7.60
Talc	203.70	8.80

Grind the above materials in a high-speed mill (Cowles, 3800–4500 fpm for 5 min) and then add:

Magnesium Aluminum Silicate	5.00	0.25

Grind for an additional 10 min and let down, at a lower speed, as follows:

Rhoplex AC-388		390.80	44.39
Long Oil Alkyd*		30.80	3.69
Proplyene Glycol	Premix	35.00	4.07
Antifoamer	Premix	1.00	0.13
Tributyl Phosphate	Premix	9.80	1.21
Ammonium Hydroxide (28%)	Premix	2.00	0.27
Water and/or Hydroxyethyl Cellulose (2.5%)		25.30	3.05
Toner: Cal/Ink **GP8814E** Phthalo Blue		0.05	–
Cal/Ink **GP8807B** Lamp Black		0.05	–

* Drier treated with 0.5% by weight of 6% cobalt, 0.5% of 6% manganese, and 1.4% of 24% lead, prior to incorporation of alkyd.

PVC (%)	40.4
Total Solids by Vol. (%)	41.2
Total Solids by Wt. (%)	58.1
pH	≈9.4-9.7
Visc.–Initial (KU)	≈77–83

No. 12

(Alkyd-Modified, High Hiding)

		lb	gal
Hydroxyethyl Cellulose (2.5% sol'n.)		30.00	3.64
Water		80.00	9.60
Tamol 731 (25%)		21.00	2.26
Triton X-405		3.00	0.33
Antifoamer		1.00	0.13
Ethylene Glycol		25.00	2.69
Propylene Glycol		35.00	4.07
Preservative		1.80	0.11
Rutile Titanium Dioxide (nonchalking)		300.00	9.12
Talc		97.70	4.22
Calcium Carbonate		78.10	3.46

Grind the above materials in a high-speed mill (Cowles, 3800–4500 fpm for 5 min) and then add:

Magnesium Aluminum Silicate		10.00	0.50

Grind for an additional 10 min and let down, at a lower speed, as follows:

Rhoplex AC-388 (50%)		400.40	45.49
Long Oil Alkyd*		31.20	3.78
Tributyl Phosphate	Premix	10.00	1.23
Antifoamer	Premix	1.00	0.13
Ammonium Hydroxide (28%)	Premix	2.00	0.27
Water and/or Hydroxyethyl Cellulose (2.5%)		74.80	8.97
Toner: Cal/Ink **GP8814E** Phthalo Blue		0.05	–
Cal/Ink **GP8807B** Lamp Black		0.05	–

* Drier treated with 0.5% by weight of 6% cobalt, 0.5% of 6% manganese, and 1.4% of 24% lead, prior to incorporation of alkyd.

PVC (%)	40.7
Total Solids by Vol. (%)	42.5
Total Solids by Wt. (%)	59.7
pH	≈9.4–9.7
Visc.—Initial (KU)	≈72–76

No. 13

(Semidripless)

	lb	gal
Hydroxyethyl Cellulose (2.5% sol'n.)	50.00	6.09
Water	90.00	10.80
Tamol 731 (25%)	15.00	1.63
Triton CF-10	2.50	0.28
Antifoamer	1.00	0.13
Ethylene Glycol	25.00	2.69
Preservative	1.80	0.11
Rutile Titanium Dioxide (nonchalking)	250.00	7.60
Talc	203.70	8.80
Magnesium Aluminum Silicate	0.50	0.02

Grind the above materials in a high-speed mill (Cowles, 3800–4500 fpm for 10–15 min) and let down, at a slower speed, as follows:

		lb	gal
Rhoplex AC-388 (50%)		459.80	52.23
Antifoamer	Premix	1.00	0.13
Tributyl Phosphate	Premix	11.50	1.41
Propylene Glycol	Premix	35.00	4.07
Ammonium Hydroxide (28%)	Premix	2.00	0.27
Water and/or Hydroxyethyl Cellulose (2.5%)		31.20	3.77
Toner: Cal/Ink **GP8814E** Phthalo Blue		0.05	–
Cal/Ink **GP8807B** Lamp Black		0.05	–

PVC (%)	40.0
Total Solids by Vol. (%)	41.0
Total Solids by Wt. (%)	58.0
pH	≈9.4–9.7
Visc.–Initial (KU)	≈76–82

No. 14

High speed mill dispersion in order of addition:

	lb/100 gal
Water	200.0
Troysan CMP-Acetate	1.0

Ethylene Glycol	25.0
Natrosol 250 HR	4.0

Mix at high speed until cellulose is completely hydrated. Reduce speed and add:

Nopco NDW	3.0
Igepal CO-630	3.0
Tamol 731	8.0
KTPP	1.5
Texanol	10.0
Water	110.0

Increase to high speed and add:

Ti Pure R-901	125.0
Glomax LL	200.0
Kaopaque 30	120.0
Silica 1160	50.0

Disperse for 10–15 min, then add:

Rhoplex AC-388	250.0
Water and/or **Natrosol 250 HR** (2% sol'n.)	60.0

PVC (%)	61.5
lb/gal	11.7
Total NV by Wt. (%)	52.6

Alkyd Interior White Finish Coat

	lb	gal
Unitane OR-650	325.3	9.77
Rezyl 387-5 (55%)	185.1	22.04

Disperse above pigment paste

	lb	gal
Rezyl 387-5 (55%)	309.4	36.83
Beetle 80	90.4	9.99
Xylene	148.4	20.97
2-Amino-2 Methyl Propanol	3.6	0.43
Catalyst 4040	1.8	0.23

Ethylene Vinyl Acetate Emulsion Paint

	Formula No. 1 lb	gal	No. 2 lb	gal
Ti Pure R-901	200.0	6.0	200.0	6.0
Drikalite	450.0	20.0	450.0	20.0
Daxad 30	5.0	0.5	5.0	0.5
Colloid 606	2.0	0.3	2.0	0.3
Ethylene Glycol	28.0	3.0	28.0	3.0
Hexylene Glycol	7.0	1.0	7.0	1.0
Troysan PMA-30	0.3	–	0.3	–
Igepal CO-610	2.0	0.2	2.0	0.2
Natrosol 250 HR (3%)	54.0	6.5	80.0	9.6
Water	238.0	28.5	211.0	25.4
Aircoflex 500 (55 NV)	310.0	34.0	–	–
Elvace PB-3 1942 (55 NV)	–	–	302.0	34.0
PVC (%)	59.7		59.8	
Visc. (KU)	80–85		80–85	
Paint NV by Vol. (%)	45.9		46.3	
Paint NV by Wt. (%)	63.5		63.6	
Vehicle NV by Wt. (%)	26.8		26.5	

Latex Interior White Flat Paint

Formula No. 1

(Dripless)

		lb	gal
Water		86.0	10.32
Tamol 731 (25%)		15.0	1.64
Triton CF-10		3.0	0.33
Hydroxyethyl Cellulose–2.5% **(QP-4400)**		50.0	6.00

Add while mixing:

		lb	gal
Colloid 600		1.0	0.13
Hexylene Glycol	Premix	30.0	3.93
Ethylene Glycol	Premix	35.0	3.77
Super Ad-It	Premix	1.0	0.12

Add:

Ti Pure R-901	180.0	5.42
Silica 1160	52.0	2.37
Icecap K	130.0	6.01
Zeolex 80	27.0	1.54

Grind on Cowles dissolver for 15 min at 4500 fpm, then add at low speed:

Rhoplex AC-22	255.0	28.82
Colloid 600	4.0	0.52
Veegum T (4% dispersion)	125.0	14.61
Hydroxyethyl Cellulose–2.5% (**QP-4400**)	97.0	11.64
Water	30.0	3.60

Adjust pH to 9.5 with ammonium hydroxide (28%).

PVC (%)	57.0
Total Solids (%)	46.3
Visc.–Initial (KU)	89

No. 2

(Good-Quality)

	lb	gal
Water	86.2	10.34
Tamol 731 (25%)	10.0	1.09
Triton CF-10	2.0	0.22
Hydroxyethyl Cellulose–2.5% (**QP-4400**)	50.0	6.05

Add while mixing:

Colloid 600		1.0	0.13
Hexylene Glycol	Premix	30.0	3.93
Ethylene Glycol	Premix	15.0	1.61
Super Ad-It	Premix	1.0	0.12

Add:

Ti Pure R-901	180.0	5.15
Silica 1160	52.5	2.39
Icecap K	130.0	6.01
Zeolex 80	27.0	1.54

Grind on high-speed dispersion equipment for 15 min, then add at low speed:

Water	43.0	5.16
Colloid 600	2.0	0.26
Rhoplex AC-22	254.7	28.78
Hydroxyethyl Cellulose–2.5% **(QP-4400)**	222.7	26.95

Adjust pH to 9.5 with ammonium hydroxide (28%).

PVC (%)	55
Total Solids (%)	46.5
Visc.–Initial (KU)	87–92

No. 3

(High-Quality, High-Hiding)

Charge the following to the tank:

	lb	gal
Water	119.0	14.38
KTPP	0.9	–
Tamol 731 (25%)	9.3	1.01
Triton CF-10	1.9	0.21
Nopco NDW	0.9	0.13
Hydroxyethyl Cellulose–2.5% **(QP-4400)**	46.6	5.64
Hexylene Glycol	28.0	3.66
Ethylene Glycol	23.3	2.50
Super Ad-It	0.9	0.11
Ti Pure R-901	256.0	7.33
Optiwhite	163.0	8.71
Celite 281	46.6	2.43
Silica 1160	37.3	1.70

Grind on high-speed dispersion equipment for 10 min then add at low speed:

Rhoplex AC-22	280.0	31.60
Nopco NDW	0.9	0.13
VM&P Naphtha	0.9	0.13
Hydroxyethyl Cellulose–2.5% **(QP-4400)**	85.0	10.29
Water	83.0	10.00

Adjust pH to 9.5 with ammonium hydroxide (28%).

PVC (%)	61
Total Solids (%)	53.3
Visc.–Initial (KU)	76

No. 4

(Economy, Contractor-Type)

		lb	gal
Water		100.0	12.00
Tamol 731 (25%)		10.0	1.09
Triton CF-10		2.0	0.22
Hydroxyethyl Cellulose–2.5% (**QP-4400**)		200.0	24.20

Add while mixing:

Colloid 600		2.0	0.26
Hexylene Glycol	Premix	25.0	3.28
Ethylene Glycol	Premix	10.0	1.08
Super Ad-It	Premix	0.5	0.06

Add:

Ti Pure R-901	75.0	2.15
Icecap K	233.0	10.76
Silica 1160	50.0	2.28

Grind on high-speed dispersion equipment for 15 min, then add at low speed:

Water	10.5	1.26
Colloid 600	2.0	0.26
Rhoplex AC-22	230.0	25.99
Hydroxyethyl Cellulose–2.5% (**QP-4400**)	133.0	16.09

Adjust pH to 9.5 with ammonium hydroxide (28%).

PVC (%)	58.2
Total Solids (%)	42.5
Visc.–Initial (KU)	88

No. 5

(High-Hiding, High Scrub)

		lb	gal
Water		128.0	15.36
Tamol 731 (25%)		10.0	1.09
Triton CF-10		2.0	0.22

Add while mixing:

		lb	gal
Colloid 600		1.0	0.13
Ethylene Glycol	Premix	15.0	1.61
Hexylene Glycol	Premix	30.0	3.93
Super Ad-It	Premix	1.0	0.12

Add:

	lb	gal
Ti Pure R-901	230.0	6.58
Icecap K	150.0	6.93
Silica 1160	52.5	2.39

Grind on high-speed dispersion equipment for 15 min, then add at low speed:

	lb	gal
Water	93.5	11.22
Colloid 600	2.0	0.26
Rhoplex AC-22	301.3	34.05
Acrysol ASE-60 (2.5% preneutralized)	128.5	15.55

Adjust pH to 9.5 with ammonium hydroxide (28%).

PVC (%)	53.4
Total Solids (%)	49.5
Visc.–Initial (KU)	73

No. 6

Charge the following to the Cowles:

	lb	gal
Propylene Glycol	70.0	8.12
Tamol 731 (25%)	11.0	1.20

Nopco NDW	2.0	0.27
Tronox CR-800	275.0	8.02

Grind at high speed for 15–20 min. Add the following at low speed:

Propylene Glycol		82.5	9.57
Rhoplex AC-490 (46.5%)		590.5	66.61
Nopco NDW	Premix	4.0	0.54
Texanol	Premix	16.5	2.08
Triton GR-7	Premix	2.0	0.23
Water		25.0	3.00
Dowicil 75		2.6	0.11
Skane M-8 (45%)		0.5	0.06
Water and/or **Natrosol 250 MR** (2½%)		50.0	6.00

Total Solids (%)	48.3
PVC (%)	22.1
Total Solids by Vol. (%)	34.2
Visc. (KU)	75–80
pH	8.5–9.0

No. 7

Charge the following to the Cowles:

		lb	gal
Propylene Glycol		70.0	8.12
Tamol 731 (25%)		11.0	1.20
Nopco NDW		2.0	0.27
Tronox CR-800		250.0	7.30
Delore X5R		50.0	1.36

Grind at high speed for 15–20 min. Add the following at low speed:

Propylene Glycol		100.0	11.60
Rhoplex AC-490 (46.5%)		492.7	55.68
Nopco NDW	Premix	2.0	0.27
Texanol	Premix	13.7	1.73
Triton GR-7	Premix	2.0	0.23

		lb	gal
Water	Premix	50.0	6.00
Dowicil 75		2.6	0.11
Skane M-8 (45%)		0.5	0.06
Water and/or **Natrosol 250 MR** (2½%)		58.3	7.00

Total Solids by Wt. (%)	47.8
Total Solids by Vol. (%)	32.2
PVC (%)	26.8
Visc. (KU)	75-80
pH	8.5-9.0

PV Acetate Interior White Paint

Add in the listed order:

	lb	gal
Water	411.00	49.32
Preservative	0.15	–
Antifoam Agent	2.00	0.31
Freeze-Thaw Stabilizer	20.00	2.15
Hydroxy Ethyl Cellulose	4.20	0.36
Dispersing Agent	9.20	1.00
Titanox C 50	475.00	16.38
Micro-Cel T-38	5.00	0.21
Calcium Carbonate	25.00	1.11
Celite 281	25.00	1.29

Grind the above in a high-speed mill and let down at a slower speed with:

	lb	gal
Wetting Agent	4.00	0.50
Coalescing Agent	8.40	1.00
Polyvinyl Acetate Copolymer	240.00	26.40

Total Vehicle (%)	56.3
Total NV Vehicle (%)	19.5
PVC (%)	58.6

Acrylic Interior Enamel

Formula No. 1

(Semigloss)

Propylene Glycol	65.0
Nopco NDW	1.0
Dowicil 75	1.0
Tamol 731	8.0
Rutile Titanium Dioxide	220.0

Mill the above in a high-speed mill (Cowles) for 20 min and let down at a slower speed as follows:

Water		93.0
Rhoplex AC-417 (48.0%)		400.0
Propylene Glycol		5.0
Acrysol G-110	Premix	10.0
Water	Premix	25.0
Texanol	Premix	12.0
Triton GR-7M	Premix	1.0
Nopco NDW	Premix	2.0
Water/**Natrosol 250 HR** (2.5% sol'n.)		183.0

PVC (%)	24.5
Total Solids by Vol. (%)	26.2
Visc.–Initial (KU)	76
pH	9.1

	No. 2 *(White, Semi-gloss)*	No. 3 *(Tint-base, Semi-gloss)*
Propylene Glycol	35.0	23.0
Water	25.0	17.0
Ammonia (28% NH_3)	3.0	2.5
Foamaster L	2.0	2.0

Mix the above at slow speed, then add the following:

Acrysol I-98 (30%)		40.0	30.0

Mix the above for 5 min at slow speed, then add the following:

Unitane OR-600		250.0	175 0
Hydrite UF		15.8	15.8

Grind the above at high speed for 20 min, then let down at a slow speed as follows:

Rhoplex AC-64 (60.5%)		397.2	376.1
Propylene Glycol	Premix	65.0	53.8
Texanol	Premix	12.0	8.6
Foamaster L	Premix	4.0	6.0
Dowicil 75	Premix	1.2	1.2
Water	Premix	25.0	25.0
Collacral VL	Premix	30.0	30.0
Water	Premix	40.0	30.0
Triton X-207		–	2.0
Triton N-57		–	4.0
Acrysol ASE-60	Premix	12.0	20.0
Water	Premix	100.8	176.9
Ammonia (28% NH_3)	Premix	4.0	3.0
PVC (%)		23.0	18.8
Total Solids by Vol. (%)		35.0	31.1
pH		9.0-9.4	9.0-9.2
Visc.–Initial (KU)		74-78	78-82

No. 4

(Semigloss)

Charge the following to the tank:

	lb	gal
Tamol 731 (25%)	7.20	0.78
Propylene Glycol	20.00	2.32
Water	30.00	3.60
Hydroxyethyl Cellulose–2½% sol'n. (**QP-4400**)	92.00	11.04
Colloids 600	2.00	0.26

Add the following at low speed:

Ti Pure R-900 HG	170.00	4.96
De Lore X5R (supermicrofine)	75.00	2.02

Grind for 20 min at high speed on a Cowles then add the following at low speed:

Rhoplex AC-388 (50%)		352.00	40.10
Colloids 600		1.00	0.13
Triton GR-7M	Premix	2.00	0.23
Super Ad-It	Premix	0.50	0.06
Water	Premix	20.00	2.40
Ethylene Glycol	Premix	20.00	2.16
Texanol	Premix	6.00	0.76
Hydroxyethyl Cellulose–2½% sol'n. **(QP-4400)*** and/or Water		243.00	29.16

* Adjust initial viscosity to about 67 KU.
Tone with Cal Ink "B" (0.4 lb/gal) to a reflectance of 82%.

PVC (%)	27.2
Total Solids by Vol. (%)	25.8
Total Solids (%)	40.8
Visc.–Initial (KU)	65-69

No. 5

(Premium Quality, Gloss)

Propylene Glycol	75.0
Tamol SG-1 (35%)	11.6
Deefo 495	2.5
Rutile Titanium Dioxide	255.0

Mill the above in a high-speed mill (Cowles) for 20 min and let down at a slower speed as follows:

Water	40.0
Rhoplex AC-417 (48%)	513.5
Acrysol RM-3 (30%)	34.3

DMAE Sol'n. (50% in water)		2.0
Deefo 495		2.5
Propylene Glycol	Premix	20.0
Butyl **Carbitol**	Premix	23.0
Texanol	Premix	7.7
Super Ad-It	Premix	1.0
Triton GR-7M	Premix	0.5
Water		7.8
Water/**Natrosol 250 MR** (2.5% sol'n.)		66.6

PVC (%)	22.0
Total Solids by Vol. (%)	34.0
Visc.—Initial (KU)	70-73
pH	8.5

Alkyd Interior Enamel

Formula No. 1

(Semigloss)

	lb	gal
Titanium Dioxide **RANC**	265.0	7.8
Vicron 15-15	230.0	10.2
Optiwhite	110.0	6.0
Thixatrol ST	3.0	0.4
Varkyd 505-70	240.0	30.0
Mineral Spirits	52.0	8.0

Intensive mixing to 135 F—obtain 6 N.S. grind.

	lb	gal
Varkyd 505-70	176.0	22.0
Mineral Spirits	130.0	20.0
Cobalt Naphthenate (6%)	1.5	0.2
Lead Naphthenate (24%)	3.5	0.4
Exkin # 2	1.0	0.1

	No. 2 (*White, Gloss*)		No. 3 (*White, Gloss*)	
	lb	gal	lb	gal
Ti Pure R-900	275.0	8.1	275.0	8.1
Kadox 515	15.0	0.3	15.0	0.3
Atomite	50.0	2.2	50.0	2.2
Cargill BB-6	540.0	68.0	—	—
Aroplaz 1266	—	—	552.0	69.0
Glyceryl-Mono-Oleate	8.0	1.0	8.0	1.0
Cobalt Naphthenate (6%)	4.0	0.5	4.0	0.5
Calcium Naphthenate (4%)	1.5	0.2	1.5	0.2
Antiskinning Agent	1.0	0.1	1.0	0.1
Mineral Spirits	128.0	19.6	121.0	18.6
PVC (%)	19.3		19.3	
Visc. (KU)	85-90		85-90	

No. 4

(*Eggshell*)

	lb	gal
Titanox C-50	500.0	17.25
Celite 281	25.0	1.30
High Acid A1. Stearate	4 0	—
Long Oil Alkyd (odorless)	360.0	48.00
Thinner (odorless)	207.0	32.81
Cobalt Naphthenate (6%)	1.8	0.23
Lead Naphthenate (24%)	2.7	0.28
Antiskinning Agent	1.0	0.13
Vehicle NV (%)	37.7	
PVC (%)	41.3	

Latex Interior Enamel

Formula No. 1

(Economy, Semigloss)

lb/100 gal

Grind the following in a high-speed mill (Cowles, 3800–4500 fpm for 15 min):

Water	60
Cellosize QP-4400 (2.5% sol'n.)	62
Tamol 731 (25%)	20
Propylene Glycol	20
Colloid 600	2
Ti Pure R-900	160
Hydrite UF	50

Let down at lower speed the following:

Rhoplex AC-490 (46.5%)	382
Texanol	10
Triton GR-7	2
Super Ad-It	5
Water	64
Ethylene Glycol	20
Colloid 600	1
Cellosize QP-4400 (2.5% sol'n.)	160

PVC (%)	28
Total Solids (%)	25
Visc. (KU)	85-90

No. 2

lb/100 gal

Tamol 731 (25%)	11
Nopco NDW	2
Water	73

Propylene Glycol	32
Ti Pure R-900	275
Dowicil 100	1
Water	20
Propylene Glycol	32
Dalpad A	10
Nopco NDW	4
Triton GR-7	2
Dowfax 2A1	2
Rhoplex AC-490 (46.5%)	590
Water	68
XD-7846.01	6

No. 3

(Semigloss, Acrylic Modified)

Anionic Dispersant (25%)	9.5
Defoamer	2.0
Propylene Glycol	100.0
Titanium Dioxide	275.0

Disperse on Cowles then add:

Acrylic Latex	550.0
Propylene Glycol	5.0
Preservative	1.0
Pine Oil	2.0
Defoamer	2.0
Water	85.0
Thickener (2½%)	60.0

No. 4

(Economy, Semigloss)

	lb	gal
Charge the following to the tank:		
Tamol 731 (25%)	7.20	0.78
Propylene Glycol	20.00	2.32

Water	30.00	3.60
Hydroxyethyl Cellulose–2½% sol'n. **(QP-4400)**	92.00	11.04
Colloids 600	2.00	0.26

Add the following at low speed:

Ti Pure R-900 HG	170.00	4.96
De Lore X5R (supermicrofine)	75.00	2.02

Grind for 20 min at high speed in a Cowles then add the following at low speed:

Rhoplex AC-490 (46.5%)		382.00	43.16
Colloids 600		1.00	0.13
Triton GR-7M	Premix	2.00	0.23
Super Ad-It	Premix	0.50	0.06
Water	Premix	20.00	2.40
Ethylene Glycol	Premix	20.00	2.16
Texanol	Premix	10.00	1.26
Hydroxyethyl Cellulose–2½% sol'n. **(QP-4400)*** and/or Water		214.00	25.68

* Adjust initial viscosity to about 67 KU.

Tone with Cal Ink "B" (0.4 lb/100 gal) to a reflectance of 82%.

PVC (%)	27.2
Total Solids by Vol. (%)	25.5
Total Solids	40.6
Visc.–Initial (KU)	65-69

No. 5

(Low-Gloss, White)

	lb	gal
Charge the following to the tank:		
Tamol 731 (25%)	15.0	1.64
Nopco NDW	4.0	0.54
Propylene Glycol	75.0	8.70

Add the following at low speed, 1900 fpm:

Ti Pure R-900	275.00	8.03
Atomite	100.0	4.43

Increase speed to 5400 fpm and grind for 20 min, then add at low speed, 1900 fpm:

Propylene Glycol	92.0	10.67
Rhoplex AC-22	420.0	47.50
Ethylene Glycol	33.0	3.55
Super Ad-It	1.0	0.12
Water	165.0	19.97
Triton GR-7	2.0	0.22
Acrysol G-110	10.0	1.19

Adjust pH to 9.5 with ammonium hydroxide (28%).

PVC (%)	39
Total Solids (%)	47.8
Visc.–Equilibrated (KU)	75-80

No. 6

(Medium-Gloss, White)

lb/100 gal

Grind the following in a high-speed mill (Cowles, 3800-4500 fpm for 15 min):

Water	15
Cellosize QP-4400 (2.5% sol'n.)	20
Tamol 731 (25%)	16
Defo-233	2
Propylene Glycol	60
Ti Pure R-900	240
Hydrite UF	20

Let down at lower speed the following:

Propylene Glycol	58
Rhoplex AC-490 (46.5%)	556
Super Ad-It	1

Texanol	16
Triton GR-7	2
Defo-233	3
Cellosize QP-4400 (2.5% sol'n.)	60

PVC (%)	23
Solids (%)	35
Visc. (KU)	85-90

Acrylic Dry-Ground Deep Tone Paint

Formula No. 1

(Gold-Yellow)

	lb	gal
Tamol 731 (25%)	9.3	1.01
Triton CF-10	2.0	0.23
Antifoamer	1.0	0.13
Ethylene Glycol	35.0	3.77
Hydroxyethyl Cellulose (2.5% sol'n.)	100.0	12.12
Water	60.0	7.20
Preservative (57% mercury)	1.8	0.11
Yellow Oxide **(YLO-1788)**	100.0	2.98
Silica	304.8	13.84

Grind the above materials in a high-speed mill (Cowles, 3800-4500 fpm for 10-15 min) and let down, at a slower speed, as follows:

		lb	gal
Rhoplex AC-388 (50%)		471.0	53.51
Long Oil Alkyd*		–	–
Antifoamer	Premix	1.0	0.13
Tributyl Phosphate	Premix	10.0	1.23
Ammonium Hydroxide (28%)	Premix	2.0	0.27
Water and/or Hydroxyethyl Cellulose (2.5%)		28.9	3.47

* Drier treated with 0.5% of 6% cobalt, 0.5% of 6% manganese, and 1.4% of 24% lead prior to incorporation of alkyd.

PVC (%)	40.0
Total Solids by Vol. (%)	42.0

Total Solids by Wt. (%)	56.8
pH	≈9.4-9.7
Visc.–Initial (KU)	≈78-84

No. 2

(Green)

	lb	gal
Tamol 731 (25%)	6.3	0.68
Triton CF-10	2.0	0.23
Antifoamer	1.0	0.13
Ethylene Glycol	35.0	3.77
Hydroxyethyl Cellulose (2.5% sol'n.)	100.0	12.12
Water	60.0	7.20
Preservative (57% mercury)	1.8	0.11
Chromium Oxide (X-1134)	100.0	2.31
Silica	318.0	14.44

Grind the above materials in a high-speed mill (Cowles, 3800–4500 fpm for 10–15 min) and let down, at a slower speed, as follows:

		lb	gal
Rhoplex AC-388 (50%)		471.0	53.51
Antifoamer	Premix	1.0	0.13
Tributyl Phosphate	Premix	10.0	1.23
Ammonium Hydroxide (28%)	Premix	2.0	0.27
Water and/or Hydroxyethyl Cellulose (2.5%)		32.3	3.87

PVC (%)	39.9
Total Solids by Vol. (%)	42.0
Total Solids by Wt. (%)	57.3
pH	≈9.4-9.7
Visc.–Initial (KU)	72-78

No. 3

(Barn Red)

	lb	gal
Tamol 731 (25%)	5.8	0.63
Triton CF-10	2.0	0.23
Antifoamer	1.0	0.13

Ethylene Glycol	35.0	3.77
Hydroxyethyl Cellulose (2.5% sol'n.)	100.0	12.12
Water	60.0	7.20
Preservative (57% mercury)	1.8	0.11
Red Iron Oxide (RO-8097)	80.0	1.96
Red Iron Oxide (R-2899)	20.0	0.46
Silica	316.7	14.38

Grind the above materials in a high-speed mill (Cowles, 3800-4500 fpm for 10-15 min) and let down at a slower speed, as follows:

Rhoplex AC-388 (50%)		471.0	53.51
Antifoamer	Premix	1.0	0.13
Tributyl Phosphate	Premix	10.0	1.23
Ammonium Hydroxide (28%)	Premix	2.0	0.27
Water and/or Hydroxyethyl Cellulose (2.5%)		32.3	3.87

PVC (%)	40.0
Total Solids by Vol. (%)	42.0
Total Solids by Wt. (%)	57.3
pH	≈ 9.4-9.7
Visc.–Initial (KU)	≈ 72-78

No. 4

(Black)

	lb	gal
Tamol 731 (25%)	6.3	0.68
Triton CF-10	2.0	0.23
Antifoamer	1.0	0.13
Ethylene Glycol	35.0	3.77
Hydroxyethyl Cellulose (2.5% sol'n.)	100.0	12.12
Water	60.0	7.20
Preservative (57% mercury)	1.8	0.11
Germantown Lamp Black	20.0	1.30
Silica	341.4	15.50

Grind the above materials in a high-speed mill (Cowles, 3800-4500 fpm for 10-15 min) and let down, at a slower speed, as follows:

Rhoplex AC-388		471.0	53.51
Antifoamer	Premix	1.0	0.13
Tributyl Phosphate	Premix	10.0	1.23
Ammonium Hydroxide (28%)	Premix	2.0	0.27
Water and/or Hydroxyethyl Cellulose (2.5%)		31.5	3.82

PVC (%)	40.0
Total Solids by Vol. (%)	42.0
Total Solids by Wt. (%)	55.1

Acrylic Interior Semigloss Tint Base for Water- and Glycol-Dispersed Colorants

Formula No. 1

(As in Cal Ink Series)

		lb	gal
Tamol 731 (25%)		10.0	1.09
Triton CF-10		1.0	0.12
Water		45.0	5.40
Balab 917		2.0	0.24
Propylene Glycol	Premix	60.0	6.96
Natrosol 250 MR (100%)	Premix	1.0	0.06

Mix the above at a slow speed for 5 min, then add:

Ti Pure R-900		200.0	5.84
Hydrite UF		30.0	1.42

Grind the above for 20 min and let down as follows:

Propylene Glycol		95.0	11.02
Rhoplex AC-388 (50%)		481.2	54.66
Water		20.0	2.40
Texanol	Premix	10.0	1.26
Super Ad-It	Premix	1.0	0.12
Balab 917	Premix	2.0	0.24
Triton N-57	Premix	4.0	0.46
Acrysol G-110 (22%)	Premix	10.0	1.80
Water	Premix	57.6	6.91

PVC (%)	22.0
Total Solids by Vol. (%)	33.0
Total Solids by Wt. (%)	46.3
Visc.–Initial (KU)	72-76

No. 2

(As in Bennett Star Colors)

		lb	gal
Tamol 850 (30%)		8.0	0.80
Propylene Glycol		60.0	6.96
Super Ad-It		1.0	0.12
Antifoamer 60		1.0	0.12
Ti Pure R-900		200.0	5.84
Barytes # 1		30.0	0.82

Grind the above for 20 min and let down as follows:

		lb	gal
Propylene Glycol		70.0	7.61
Rhoplex AC-388 (50%)		492.4	55.94
Texanol	Premix	10.0	1.26
Water	Premix	20.0	2.40
Triton X-207	Premix	10.0	1.22
Triton X-114	Premix	1.5	0.17
Acrysol G-110 (22%)	Premix	15.0	1.80
Water	Premix	124.5	14.94

PVC (%)	20.2
Total Solids by Vol. (%)	33.0
Total Solids by Wt. (%)	45.6
Visc.–Initial (KU)	72-76

Acrylic Economy Tint Base

Formula No. 1

Natrosol 250 MHR (100%)	2.5
Ethylene Glycol	30.0
Water	150.0
Tamol 850 (30%)	6.9

KTPP	1.5
Triton CF-10	1.0
Foamaster VL	1.0
Ti Pure R-902	100.0
AZO-11	25.0
Snowflake	302.3
Mica 325	20.0

Grind for 5 min, then add:

Attagel 50	10.0

Let down:

Rhoplex AC-64 (60.5%)	259.5
Texanol	7.7
Foamaster VL	3.0
Skane M-8	1.0
Water and/or **Natrosol MHR** (2.5%)	241.2

PVC (%)	52.0
Total Solids by Vol. (%)	35.0
pH	9.5
Visc. (KU)	92-98

No. 2

Natrosol 250 MHR (2½%)	150.0
Water	101.2
Tamol 850 (30%)	5.4
KTPP	0.5
Triton CF-10	1.0
Foamaster VL	1.0
Ethylene Glycol	25.0
Ti Pure R-902	100.0
AZO-11	25.0
Snowflake	221.0

Grind for 5 min, then add:

Attagel 50	10.0

Let down:

Rhoplex AC-64 (60.5%)		73.7
Texanol		5.1
Skane M-8		1.0
Foamaster VL		3.0
Triton GR-5M		1.0
Water and/or **Natrosol 250 MHR** (2.5%)		264.2
PVC (%)	55	
Total Solids by Vol. (%)	25	
pH	9.5	
Visc. (KU)	92-98	

Acrylic Interior Flat Universal Tint Base

Formula No. 1

	lb	gal
Charge the following to the tank:		
Water	60.0	7.20
Hydroxyethyl Cellulose (3.5% sol'n.)	100.0	12.00
Tamol 850 (30%)	4.0	0.40
KTPP	1.3	–
Ethylene Glycol	25.0	2.69
Hexylene Glycol	20.0	2.62
Preservative (10% mercury)	1.0	0.13
G.E. Antifoam 66	0.5	0.06
Add while mixing at low speed, 1900 fpm:		
Rutile Titanium Dioxide (nonchalking)	143.0	4.30
Icecap K	35.0	1.62
Kaopaque 10	80.0	3.70
Silica 1160	100.0	4.55

Grind the above on a high-speed disperser for 15 min at 4500 fpm; then let down the following at low speed, 1900 fpm:

Rhoplex AC-388 (50%)	271.5	30.95
Water	75.0	9.00

Hydroxyethyl Cellulose (3.5% sol'n.)	186.0	22.32
Triton N-57	12.0	1.40

Adjust pH to 9.5 with ammonium hydroxide (28%).

PVC (%)	49.5
Total Solids (%)	44.25
Visc.–Initial (KU)	≈78

Acrylic Interior Eggshell Tint Base

Formula No. 1

Propylene Glycol	50.0
Tamol 731 (25%)	10.0
Water	50.0
Dowicil 75	1.0
Natrosol 250 GR (100%)	2.0
Triton CF-10	2.0
PAG-188	2.0
Zopaque RCL-9	250.0
Imsil A-15	75.0

Grind the above in a high-speed mill (Cowles) for 20 min and let down at a slower speed as follows:

Water		50.0
Rhoplex AC-417 (48%)		422.0
Texanol	Premix	15.0
Propylene Glycol	Premix	13.0
PAG-188		4.0
Triton N-57		4.0
Acrysol RM-4 (30%)	Premix	14.0
Water	Premix	26.3
Ammonium Hydroxide	Premix	3.4
Water/**Natrosol 250 GR** (4% sol'n.)		103.2

PVC (%)	34.0
Total Solids by Vol. (%)	31.6
Visc.–Initial (KU)	80-85
pH	8.5-9.0

Latex Tint Base

Formula No. 1

(Universal)

	lb	gal
Charge the following to the tank:		
Water	60.0	7.20
Hydroxyethyl Cellulose–3.5% (**QP-300**)	100.0	12.00
Tamol 850 (30%)	4.0	0.40
KTPP	1.3	–
Ethylene Glycol	25.0	2.69
Hexylene Glycol	20.0	2.62
Super Ad-It	1.0	0.13
G.E. Antifoam 66	0.5	0.06
Add at low speed, 1900 fpm:		
Ti Pure R-901	143.0	4.30
Icecap K	35.0	1.62
Kaopaque 10	80.0	3.70
Silica 1160	100.0	4.55
Increase speed to 4500 fpm and grind for 10 min; then let down at low speed, 1900 fpm:		
Rhoplex AC-22	327.0	36.95
Water	73.0	8.76
Hydroxyethyl Cellulose–3.5% (**QP-300**)	137.0	16.44
Triton N-57	12.0	1.40

Adjust pH to 9.5 with ammonium hydroxide (28%).

PVC (%)	48.4
Total Solids	47.1
Visc.–Initial (KU)	79

No. 2

(White and Light)

	lb	gal
Pigment grind:		
Water	292.0	35.0
Nopcosant L	8.0	0.8
KTPP	1.5	0.2
Merbac 35	0.5	–
Ethylene Glycol	25.0	2.6
Igepal CO-630	3.0	0.3
Colloid 677	1.5	0.2
Cellosize QP-4400H	4.0	0.4
Ti Pure R-901	200.0	6.0
Micro-Cel T-70	40.0	2.1
Celite 281	40.0	2.1
Al-Sil-Ate NC	100.0	4.6
Atomite	75.0	3.3
Ucar Filmer 351	10.0	1.3
Let down:		
Ucar Latex 365	180.0	19.8
Water	176.0	21.1
Colloid 677	2.0	0.2

PVC (%)	63.5
Total Solids (%)	48.7
Visc. (KU)	80-90
lb/gal	11.6

No. 3

(White and Light)

	lb	gal
Dispersion		
Water	300.0	36.00
Natrosol 250 HR	4.0	0.35
Daxad 30 (25%)	6.0	0.63

KTPP	0.5	0.02
PMA-18	0.3	0.03
Igepal CO-630	4.0	0.46
Ethylene Glycol	25.0	2.70
Titanox RA-47	250.0	7.32
Micro-Cel T-70	20.0	1.06
Snowflake White	100.0	4.43
ASP-400	125.0	5.82
Reduction:		
Texanol	16.0	2.02
Everflex E (55%)	250.0	27.78
Nopco NDW	2.0	0.27
Water	95.8	11.50

PVC (%)	56.8
Total Solids %)	53.3
Total Solids by Vol. (%)	33.3
Visc. (KU)	85-90

No. 4

(White and Light)

	lb	gal
Charge into mixer under agitation:		
Water	160.0	19.2
KTPP	1.0	0.1
Igepal CTA-639	4.0	0.4
PMA-30	0.5	–
Tamol 731	4.0	0.4
Ethylene Glycol	25.0	2.6
Titanox RA-51	220.0	6.4
Micro-Cel T-70	30.0	1.6
ASP-400	50.0	2.3
Atomite	125.0	5.5
Colloid 677	1.0	0.1
Natrosol 250 (1.5%)	50.0	6.0
Carbitol Acetate	10.0	1.2

Disperse in high-speed mill, then add slowly with agitation:

Latex Copolymer (55% solids)	260.0	28.6
Flexbond 860		
Resyn 2243		
Colloid 677	2.0	0.2
Natrosol 250 HR (1.5%) or Water	210.0	25.2
Total Solids (%)	50.2	
PVC (%)	52	
lb/gal	11.6	

No. 5

(White and Light)

Natrosol 250 HR (2% sol'n.)		250.0
R&R 551		3.0
Igepal CO-610		4.0
Tamol 850		8.0
Ethylene Glycol		30.0
Nopco NDW		2.0
Phenylmercuric Acetate (18% mercury)		0.3
Titanox RA-46		200.0
Celite 281		25.0
Satintone # 1		50.0
Snowflake White		200.0
Water		163.0
Aircoflex		240.0
Total Solids (%)	52	
PVC (%)	57	

No. 6

(White and Light)

Natrosol 250 HR (2% sol'n.)	300.0
R&R 551	3.0
Igepal CO-610	3.0
Tamol 850	5.0

Ethylene Glycol	20.0
Colloid 677	2.0
Phenylmercuric Acetate (18% mercury)	0.3
Titanox RA-46	150.0
Micro-Cel T-70	30.0
Satintone #1	125.0
Camel Carb	50.0
Gold Bond R Silica	75.0
Water	208.0
Aircoflex 500	165.0
Total Solids (%)	46
PVC (%)	65

No. 7

(Semigloss)

Water	28.0
Propylene Glycol	54.0
Tamol 731	9.0
Dapro 881-S	2.0
Du Pont R-900	260.0
Barytes	15.0
Natrosol 250 MR	0.3
Ammonium Hydroxide	1.0

Mix on a high-speed mill and add:

Propylene Glycol	84.0
Water	22.0
Texanol	15.0
Super Ad-It	1.0
Par Cryl 400	473.0
Water	83.0
Acrysol G-110 (as needed)	8.0
PVC (%)	24
Total Solids (%)	49
Visc. (KU)	68-72

No. 8

(Semigloss)

		lb/100 gal
Charge the following to the tank in order of addition:		
Zopaque RCS-9 (76.5% solids)		288
Kaopaque 10 (67.6% solids)		74
Propylene Glycol		167
Nopco NDS		2
Rhoplex AC-490 (46.5% solids)		488
Super Ad-It	Premix before adding	1
Butyl **Cellosolve**		23
Igepal CO-610		8
Natrosol 250 HR (2% sol'n.)		43

PVC (%)	27
lb/gal	10.9
NV by Wt. (%)	46
Visc.—Sheared (KU)	78

No. 9

(One-Coat, PVAc-Modified, White Flat)

	lb	gal
Water	145.0	17.5
Strodex PK-90	2.0	0.2
Strodex SEK-50	10.0	1.2
Troy 666	1.5	0.2
Preservative	—	—
Ethylene Glycol	25.0	2.6
Texanol	10.0	1.2
Micro-Cel T-70	40.0	2.1
Celite 281	25.0	1.3
Rutile Titanium Dioxide (high absorption grade)*	250.0	7.3
Optiwhite or **Altowhite**	75.0	4.4
Camel-Carb	100.0	4.4

Disperse in high-speed disperser, then add slowly with agitation:

Hydroxyethyl Cellulose–2% sol'n. (4400 cps)	150.0	18.0
PVAc Copolymer Resin Emulsion (55% NV)	244.0	26.8
Water and/or Hydroxyethyl Cellulose (2% sol'n.)	120.5	14.4
Troy 666	0.5	–

* **Titanox 2030, Ti Pure R-901, Zopaque R-770,** or equivalent.

Visc. (KU)	80-85
PVC (%)	58.5
Total Solids by Wt. (%)	53.5

Vinyl Acrylic Flat Tint Base

(White and Light)

	lb	gal
Charge into mixer under agitation:		
Water	85.0	10.2
KTPP	1.0	0.1
Tamol 731	4.5	0.4
Ethylene Glycol	30.0	3.2
Colloid 677	1.0	0.1
Cosan PMA-30	0.5	–
Advawet 33	4.0	0.5
Micro-Cel T-70	20.0	1.1
Atomite	150.0	6.7
Natrosol 250 HR (1½% sol'n.)	130.0	15.4
Ti Pure R-901	200.0	5.9
Gold Bond R	75.0	3.4
Disperse in high-speed mill, then add slowly with agitation:		
Resyn 2345	280.0	30.8
Texanol (add slowly)	12.0	1.5
Natrosol 250 HR (1½% sol'n.)	175.0	20.7
Colloid 677	1.5	0.2

Note:
Protect thickener solution with 0.08% **PMA 30** if stored prior to use.

PVC (%)	52
lb/gal	11.7
Total Solids (%)	52.5
Visc. (KU)	90-95

Styrene-Acrylic Interior Textured Finish

Formula No. 1

(Washable Matte)

Vinacryl 7172	140
Bermocoll E320G (5% aq. sol'n.)	70
Vinapol 1640 (diluted 2:5 with water)	35
Butyl Dioxitol	20
Propylene Glycol	10
Proxel CRL	1
Tioxide R-CR2	30
Fordacal M2	372
Fordacal SD400	300
Water	10
Paraplex WP1	10
Foamaster NDW	2

No. 2

(Economy, Matte)

Vinacryl 7172	80
Bermocoll E320G (5% aq. sol'n.)	75
Vinapol 1640 (diluted 2:5 with water)	40
Butyl Dioxitol	20
Propylene Glycol	10
Proxel CRL	1
Tioxide R-CR2	30
Fordacal M2	392
Fordacal SD400	310

Water	30
Paraplex WP1	10
Foamaster NDW	2

No. 3

(Brush Applied, Interior-Exterior)

Mix:

Vinacryl 7172	250
Bermocoll E320G (5% aq. sol'n.)	60
Vinapol 1640	6
Water	14
Propylene Glycol	10
Butyl Dioxitol	10

Then add in the order shown:

Tioxide R-CR2	70
Omya D5	290
Omya D130	220
Granicalcium (0.3-0.7 mm)	62
Proxel AB	1
Foamaster NDW	2

When fully dispersed, add:

Viscalex HV30	5

No. 4

Calgon S (5% aq. sol'n.)	9
Natrosol 250 HR (2% aq. sol'n.)	115
Acticide THP	1
Butyl Dioxitol	11
Potassium Hydroxide (50% aq. sol'n.)	1
Vinacryl 7172	137
Foamaster NDW	1
Tioxide R-CR2	69
Microdol 1	230

Microdol 40/200	343
Granicalcium (0.3–0.7 mm)	80
Viscalex HV30	8

No. 5

(Washable and Flexible)

Premix:

Bermocoll E320G (5% aq. sol'n.)	60
Water	21
Vinapol 1640	9
Propylene Glycol	10
Proxel CRL	1

Add and disperse:

Omya D5	290
Omya D130	229
Tioxide R-CR2	70

Then add:

Vinacryl 7172*	250
Granicalcium (0.3–0.7 mm)	30
Wood Chips (1200M granular)	8
Paraplex WP1	20
Foamaster NXZ	2

* Some emulsion may need to be added with the pigment and fillers to obtain workability.

Note:

Capable of withstanding movement; for use on ceilings or walls; washable.

Vinyl Acetate Textured Paint

Kronos RN44	75
Tioxide A-HR	25
Microcal ET	10

Celacose CP10 (4% aq. sol'n.)	20
Vinapol 1640 (diluted 1:6 with water)	20
Sodium Benzoate (solid)	3
Potassium Hydroxide (50% aq. sol'n.)	2
Dalpad A	10
Acticide APA	3
Foamaster NXZ	1
Water	35
Vinamul 6705	257
Dolomite AC	539

P/B Ratio	5:1
NV (%)	78
pH	7.8

Note:

The formula gives excellent exterior durability, and provides one coat cover; it bridges small blowholes and other minor surface defects in cast concrete, and can be stippled or left as an even finish. **Tioxide A-HR** is included to induce controlled chalking and the maintenance of a clean white appearance on weathering; it is unnecessary if the paint is to be tinted.

Latex Interior-Exterior Emulsion Paint

	lb	gal
Water	150	18.0
Tamol	7	0.8
Triton X-102	4	0.5
Ethylene Glycol	20	2.1
Foamicide and Wetting Agent **KP-140**	4	0.5
Vicron 15-15	75	3.3
Ti Pure R-901	200	5.7
Fibrene C-500	50	2.2
Gold Bond R	75	3.3
Cellosize QP-15000 (2.5% sol'n.)	100	12.0
Metasol 57 (liquid)	3	0.4

Mix and roller mill:

Cellosize QP-15000 (2.5% sol'n.)	100	12.0
Carbitol Solvent	13	1.5
Varaqua 928	310	34.0
Water	50	6.0

PVC (%)	43
Total NV by Wt. (%)	50.1
lb/gal	11.35
Visc. (KU)	85-90

Chapter VII

VARNISHES, LACQUERS, AND WOOD FINISHES

Varnishes and lacquers satisfy the demand for a clear finish that protects, and in the case of wood, displays the beauty of the substrate. Exterior wood varnishes easily prevent weathering, but must also prevent ultraviolet light degradation of the wood surface because the resultant deterioration causes a loss of film adhesion. Generally dark-colored (UV absorbent) products are more successful.

Butyrate-Acrylic Wood Lacquers

	Formula No. 1	No. 2	No. 3
CAB-381-0.5	8.50	8.50	8.30
Acryloid B-66 (40%)	21.30	17.30	20.70
Santicizer 160	3.00	3.00	3.00
DC-510 (1000 cs)	0.01	0.01	0.01
Eastman Inhibitor DOBP	0.09	0.09	0.09
Toluene	32.10	34.50	31.20
Tecsol 3 (95%)	10.00	10.00	9.70
Ethyl Acetate	5.00	5.00	4.80
Isobutyl Acetate	10.00	10.00	9.70
MIAK	10.00	10.00	9.70
SAIB	–	1.60	2.80
NV by Wt. (%)	20.1	20.1	22.5
Visc., Brookfield, 25 C (cps)	50	40	50
lb/gal	7.54	7.54	7.54
kg/liter	0.903	0.903	0.903
lb/Imperial gal	9.05	9.05	9.05

Acrylic Aqueous Lacquer

Formula No. 1

(White)

	lb	gal
Cowles grind. Add in order under agitation:		
Rhoplex WL-91 (41.5%)	160.0	18.56
Tamol 165	12.3	1.43
Triton CF-10	1.6	0.16
Deefo 806-10002	2.3	0.31

Sift in slowly:

Zopaque RCL-9	182.0	5.33

Disperse under high-speed Cowles for 20 min to 7½ (N.S.) grind. Let down with the following under mild agitation:

Rhoplex WL-91	495.4	57.47
Propasol DM	79.6	9.95
Aromatic 150	11.4	1.54
Dibutyl Phthalate	13.6	1.55
BYK 301	1.4	0.17
Ammonium Benzoate (10% sol'n.)	28.4	3.38

Adjust to final spray viscosity with 15% ammonium and/or **Aromatic 150.** Filter.

pH	7.05-8.0
Visc. (s–#4 Ford Cup)	27-30
Total Solids by Wt. (%)	47.2
Total Solids by Vol. (%)	38.4
P/B Ratio	40:60
PVC (%)	15

No. 2

(Green)

	lb	gal
Pebble mill grind and charge the following:		
Tamol 165	1.8	0.21
Triton CF-10	0.3	0.03
Deefo 806-102	0.3	0.04
Water	23.2	2.78
Butyl **Cellosolve**	16.3	2.17
Chrome Oxide **X-1134**	32.4	0.74
Imperial Brazil Yellow **X286C**	5.3	0.45
Phthalo Blue **417D**	0.3	0.02
Pebble mill to 7½ NS		
Shake with:		
Water	12.6	1.52
Mill 30 min, recheck grind, drain.		
Add mill base to **Rhoplex WL-91**, *slowly*, with good agitation.		
Rhoplex WL-91	606.6	70.37
Add in order under good agitation to above:		
Water	42.2	5.06
Ammonium Benzoate (10%)	25.3	3.01
Butyl **Carbitol**	25.3	3.18
Butyl **Cellosolve**	50.6	6.75
Paint Additive # 14	1.1	0.14
BYK 301	1.1	0.14
Deefo 806-102	1.1	0.14
Adjust to pH of 8.6 with:		
Ammonium Hydroxide (5%)	10.5	1.25
Reduce for spray with:		
Water	16.0	2.00

Filter.

pH	8.6
Visc. (s—#4 Ford Cup)	27-29
Total Solids by Wt. (%)	32.8
Total Solids by Vol. (%)	29.0
lb/gal	8.71
P/B Ratio	13:87
PVC (%)	4.2

No. 3

(Black)

	lb	gal
Premix:		
Acrysol I-94 (30%)	14.6	1.70
Water	23.1	2.77

Add with good agitation in order:

	lb	gal
Ammonium Hydroxide (28%)	1.1	0.12
Butyl **Cellosolve**	23.1	3.08
Raven 450	8.1	0.55

Pebble mill to 7½+ (N.S.). Drain.
Add the mill base to the following mixture of **Rhoplex WL-91** and water *slowly* with good application:

	lb	gal
Rhoplex WL-91 (41.5%)	601.0	69.71
Water	73.2	8.78

Add in order with good agitation:

	lb	gal
Ammonium Benzoate (10%)	25.1	2.99
Butyl **Cellosolve**	67.9	9.05
Texanol	10.0	1.26
BYK 301	1.0	0.13

Adjust to pH of 8.0 with:

	lb	gal
Ammonium Hydroxide (14%)	2.1	0.25

Reduce to spray viscosity with water. Filter.

Black Acrylic Lacquer

Formula No. 1

Sand mill grind:

Excelsior Black	7.5
Acryloid B-99 (50%)	15.0
Xylene/**Cellosolve** Acetate (3/1)	77.5

Lacquer let down:

Pigment Grind	6.7
Acryloid B-48N (45%)	34.2
Xylene/**Cellosolve** Acetate (3/1)	59.1

Total Solids (%)	16.5
P/B Ratio	3:97
Visc. (s—#4 Ford Cup)	≈ 16

No. 2

Ball mill grind:

Neospectra Beads AG	7.0
Acryloid B-99 (40%)	50.0
Xylene/**Cellosolve** Acetate (3/1)	43.0

Lacquer let down:

Pigment Grind	7.1
Acryloid B-44 (40%)	35.6
Xylene/**Cellosolve** Acetate (3/1)	57.3

Add 0.05% **Raybo 3** based on total lacquer.

Total Solids (%)	16.5
P/B Ratio	3:97
Visc. (s—#4 Ford Cup)	≈ 16

Thermoplastic Air-Dry Acrylic/Butyrate Wood Lacquer

Formula	No. 1	No. 2
CAB-531-1	5.5	5.5
CAB-551-0.2	3.0	3.0
Thermoplastic Acrylic Resin (100%)	8.5	8.5
Butyl Benzyl Phthalate	3.0	3.0
Toluene	14.4	40.5
Naphtha T*	25.5	–
Isopropyl Alcohol	8.4	10.7
Isobutyl Alcohol	6.0	5.3
Isobutyl Isobutyrate (IBIB)	5.3	4.1
Methyl Ethyl Ketone (MEK)	13.0	12.0
Isobutyl Acetate	5.3	5.3
Ektasolve EE	2.1	2.1
Total Solids by Wt. (%)	20	20
Visc. (cps)	61	57

* Pseudonym for any aliphatic naphtha with an evaporation rate similar to that of toluene.

Gloss White Flexible Butyrate Lacquer

(Alkyd Modified)

	lb	gal
Ti Pure R-900	150.0	4.42
M-P-A 60 (xylene)	2.9	0.40
Cellosolve Acetate	8.1	1.00
Stepan Alkyd 4209	60.2	7.00

Three roll mill grind, 3 passes.

		lb	gal
Stepan Alkyd 4209		238.3	27.74
Half Second Butyrate	Predissolve	49.8	4.98
Monochlorobenzene	Predissolve	79.1	8.56
Toluene	Predissolve	70.3	9.72
Ethanol (anhydrous)	Predissolve	49.8	7.32
SF-69 (2% in **Cellosolve** Acetate)		2.0	0.25

Chevron #225	64.4	10.30
Ethanol (anhydrous)	47.4	6.97
Monochlorobenzene	132.1	14.29
Total Solids by Wt. (%)	36.6	
Pigment/Resin Ratio	43:57	
Visc. (s–#4 Ford at 25 C)	77.0	

Pigmented Nitrocellulose Lacquer

(Alkyd Modified)

	lb/100 gal
Grind (ball mill, 110 F):	
Titanox RA-50	52
Asbestine 3X	48
M-P-A 60 (xylene)	0 (A) & 16
Aroplaz 2477 X-65	65
Estynox 140	18
½ s Nitrocellulose Sol'n.	200
Ethanol	45
Ethyl Acetate	135
Toluene	275

Nitrocellulose Lacquer

Formula No. 1

½ s RS Nitrocellulose	40
Duraplex ND-77B	35
Paraplex RG-2	20
Monoplex DBS	5

No. 2

Titanium Dioxide	25.00
½ s RS Nitrocellulose	18.75
Duraplex A-27 (drier-treated)	41.25
Paraplex RG-2	15.00

	No. 3	No. 4
½ s RS Nitrocellulose	8	8
Maleic Resin	4	4
Plasticizer	4	4
Methyl Ethyl Ketone	8	8
MIAK (methyl isoamyl ketone)	5	5
Tecsol 3 (95%)	5	5
Isobutyl Alcohol	3	3
Isobutyl Acetate	12	12
Toluene	47	47
Nonoxidizing Coconut Alkyd (100%)	4	–
SAIB	–	4
Total NV by Wt. (%)	20	20
Brookfield Visc.–cps (25 C)	52	30
lb/gal (25 C)	7.60	7.56

Butyrate-Amino Paper Lacquers

Formula	No. 1	No. 2	No. 3
CAB-381-0.5	14.50	11.00	13.20
Urea-Formaldehyde (60%) Resin	24.00	10.50	20.00
DC-510 (1000 cs) Fluid	0.05	0.05	0.05
Dow 276-V9	10.00	7.00	9.10
Epon 812	2.00	2.00	1.80
TY-3 Catalyst	0.60	0.60	0.60
SAIB	–	11.00	9.00
Isobutyl Acetate	15.50	15.50	12.55
Tecsol 3 (anhydrous)	19.70	19.70	21.20
Toluene	13.65	22.65	12.50
NV by Wt. (%)	37.6	37.6	45.5
Brookfield Visc.–cps (25 C)	645	200	695
lb/gal	7.95	7.91	8.10

Cellulose Nitrate Paper Lacquer

Formula	No. 1	No. 2	No. 3
½ s RS Nitrocellulose	16.50	16.50	15.00
Aroplaz 930	4.75	–	4.30
Kodaflex NP-10	3.75	3.75	3.40

SAIB	–	4.75	9.00
Acetone	15.00	15.00	13.50
MIBK (methyl isobutyl ketone)	10.00	10.00	9.00
Tecsol 3 (95%)	13.00	13.00	12.00
Toluene	37.00	37.00	33.80
NV by Wt. (%)	25.0	25.0	31.7
Brookfield Visc.–cps (25 C)	270	265	281
lb/gal	7.70	7.75	7.80

Alkyd Overprint Varnish

Formula	No. 1	No. 2	No. 3
Duraplex ND-78 (60%)	42	30.3	38.2
Uformite MX-61 (50%)	18	18.0	16.4
p-Toluenesulfonic Acid	1	1.0	1.0
SAIB	–	7.0	9.0
Ethyl Acetate	7	7.0	6.4
Isobutyl Acetate	5	5.0	4.5
Tecsol 3 (anhydrous)	7	7.0	6.4
Toluene	20	24.7	18.1
NV by Wt. (%)	35.0	34.2	40.1
Brookfield Visc.–cps (25 C)	25	16	26
lb/gal	8.00	8.01	8.12

Epoxy Ester Floor Varnish

Linseed Fatty Acids	9.70
Soya Fatty Acids	29.64
W.G. Rosin	18.46
Sodium Benzoate	0.05
Epoxy Resin **520-C**	21.07
Epoxy Resin **530-C**	21.08

Acrylic Tradesales Brushing Varnish

Formula	No. 1	No. 2	No. 3	No. 4	No. 5	No. 6	No. 7
Neo Cryl A603A	100	100	100	100	100	100	100
Isopropanol	5	—	—	—	—	—	—
Cellosolve	5	—	—	—	—	5	5
Texanol	—	10	—	—	—	—	5
Carbitol	—	—	10	—	5	—	—
Propylene Glycol	—	—	—	10	—	5	—
Butyl **Cellosolve**	—	—	—	—	5	—	—
Visc. (cps)	75	1075	50	55	130	55	120
pH	7.6	7.2	7.6	7.6	7.6	7.6	7.6

Acrylic Wood Finish

Formula	No. 1 (Stain)	No. 2 (Sealer)	No. 3 (Top-coat)
NeoCryl A-620	665.0	296.9	850.0
Water	35.0	15.6	150.0
Dispersant	2.0	–	–
Raw Umber Dispersion	70.0	–	–
Propylene Glycol	140.0	–	–
Zinc Stearate	–	5.0	–
Visc. (cps)	65	260	25
pH	8.5	8.4	7.9

	No. 4 (Stain)	No. 5 (Sealer)	No. 6 (Top-coat)
NeoCryl A-601	769.2	615.0	–
NeoCryl A-604	–	–	975.6
Raw Umber	76.9	–	–
Black Dye	0.4	–	–
Propylene Glycol	153.8	–	–
Zinc Stearate	–	10.0	–
Flatting Agent	–	–	24.4
Water	–	312.5	–
Ammonium Hydroxide (28%)	–	31.3	–
Butyl **Cellosolve**	–	31.3	–
Visc.	40	20	1000
pH	7.8	10.4	7.8

Alkyd Modified Flexible Polyester Wood Coating

Formula	No. 1	No. 2
A **Polylite 32-702**	40.0	36.0
Polylite 32-703	20.0	18.0
Styrene (monomer)	35.0	35.0

Cobalt Naphthenate	0.3	0.3
CAB-381-0.5	4.0	4.0
SAIB	–	6.0
B MEK Peroxide	0.7	0.7
NV by Wt.–Cured (%)	100.0	100.0
lb/gal	8.50	8.52

Catalyzed Wood Topcoat

	lb	gal
Arolon 585-W-43	339.2	38.11
Butyl **Cellosolve**	6.6	0.89
Mobilcer M	6.6	0.81
Beetle 65 (100%)	214.2	21.42
Triton X-405	7.3	0.79
Polyco 296W	18.3	1.83
Water	171.3	20.57
Second package (added at time of use):		
Catalyst 1010	113.7	15.57
NV After Catalyst (%)	46.5	
Visc. (s–#4 Ford Cup) (with catalyst)	25-30	

Polyester, Clear, Force-Drying Wood Finish

	lb/100 gal
Alcohol Soluble Butyrate (100%)	39.4
Cyplex 1473-5 (65%)	123.2
Cymel	53.3
Dow-Corning Paint Additive No. 1	0.9
Ethylene Glycol Monoethyl Ether Acetate	26.0
Methyl Ethyl Ketone	52.1
Methyl Isobutyl Ketone	156.3
Xylol	115.5
Toluene	105.8
Butanol	71.4

Catalyst 1010—add at time of use		8.9
NV by Wt. (%)	23.3	
Visc. (s—#4 Ford Cup)	17	

Unsaturated Polyester Wood Coating

	Formula No. 1	No. 2	No. 3
A Unsaturated Polyester Resin	60.0	54.0	51.0
Styrene (monomer)	34.7	34.7	34.7
Cobalt Naphthenate	0.3	0.3	0.3
CAB-381-0.5	4.0	4.0	4.0
SAIB	—	6.0	9.0
B MEK Peroxide	1.0	1.0	1.0
NV by Wt.—Cured (%)	100	100	100
Brookfield Visc.—cps (25 C)	175	170	170
lb/gal	8.61	8.67	8.69

Vinyl Acrylic Wood Finish for Spray

Formula No. 1

(Satin, Air or Force-Drying)

Mix together and disperse:

Vinacryl 4039	
Ammonia Solution (≈ 0.880 sp. gr.)	3.5
Water	116.0
Degussa TS100	20.0

Add slowly with stirring:

Texanol	54.5

Then add with stirring:

Bevaloid 677		1.0
Acticide BG		1.0
Polymul MS40		25.0
NV (%)	≈ 34	
pH	7.5-8.0	
Visc. (s–20 C, BS B4 Flow Cup)	28-33	

No. 2

(Glossy, Air or Force-Drying)

Mix together:

Vinacryl 4039	833.0
Ammonia Sol'n. (≈ 0.880 sp. gr.)	3.8
Water	78.2

Add slowly with stirring:

Texanol	58.0

Then add with stirring:

Bevaloid 677	1.0
Acticide BG	1.0
Polymul MS40	25.0

Add to the above quantity immediately before use:

XAMA-7		20.0
Total NV (%)	≈ 36	

Chapter VIII

SPECIALTY PAINTS

These formulas are representative of an expanding demand for coatings that possess unique characteristics.

Polyvinyl Chloride Flame Retardant Paint

Formula No. 1

	lb	gal
Under agitation, add the following:		
Water	102.19	12.25
Cellosize QP4400 (2%)	92.18	11.00
Ethylene Glycol	25.00	2.07
KTPP	2.50	0.13
Tamol 731 (25%)	18.00	1.97
Sodium Hexametaphosphate	11.00	1.02
Butyl **Carbitol** Acetate	7.74	0.95
Defoamer	2.50	0.36
Cosan 171-S	0.20	0.02
Mix well, then add under agitation:		
Titanium Dioxide	125.00	3.66
Alsilate NC-01	100.00	4.65
Nytal 300	218.00	9.21
Calcium Carbonate	207.00	9.21

Agitate at high speed 10–15 min; reduce speed, add, thin down under agitation:

Defoamer	2.50	0.36
Igepal CO-897	11.00	1.25
Polidene P-761	350.00	36.00

Hold for control

Water	25.00	3.00
Cellosize QP4400 (2%)	25.14	3.00

lb/gal	13.0-13.4
pH	8-9
Visc. (KU)	80-85
Total NV by Wt. (%)	63.7
Total NV by Vol. (%)	43.89
PVC (%)	61.0

No. 2

	lb	gal
Charge the following under agitation:		
Water	137.45	16.48
Cellosize QP4400 (2%)	117.32	14.00
Ethylene Glycol	25.00	2.07
KTPP	2.50	0.02
Tamol 731 (25%)	18.00	1.97
Sodium Hexametaphosphate	11.00	1.02
Butyl **Carbitol** Acetate	7.74	0.95
Defoamer	2.50	0.36
Mix well, then add:		
Titanium Dioxide	125.00	3.66
Busan 11-M-1	50.00	1.80
Alsilate NC-01	100.00	4.65
Nytal 300	218.00	9.21
Calcium Carbonate	72.00	3.20
Agitate 10-15 min at high speed, lower speed, thin under agitation:		
Water	25.00	3.00
Defoamer	2.50	0.36

Igepal CO-897	11.00	1.25
Polidene P-761	350.00	36.00

lb/gal	12.55-12.95
pH	8.5-9.0
Stabilized Visc. (KU)	70-75
Total NV by Wt. (%)	61.00
Total NV by Vol. (%)	39.52
PVC (%)	57.00

No. 3

(Intumescent)

	lb	gal
Water	187.75	22.51
KTPP	10.00	0.50
Surfactol 365	4.00	0.40
Colloids 691	3.00	0.30
Sodium Hexametaphosphate (10% sol'n.)	5.60	0.63
Cosan PMA-30	1.00	0.10
Ethylene Glycol	46.40	5.00

Mix well, thin under agitation:

Titanium Dioxide	60.00	1.80
Phoscheck P-30	250.00	17.30
Dipentaerythritol	119.70	10.22

Grind at high speed 5-10 min; lower speed, thin under slow agitation:

Unichlor 7065	43.00	3.66
Hercules X-158 (1% sol'n.)	41.58	4.96
Polidene P-761	340.28	34.90

lb/gal	10.8-11.2
Visc. (KU)	85-90
Total NV by Wt. (%)	61.4
Total NV by Vol. (%)	45.7
PVC (%)	36.2

Acrylic Automotive Clear Topcoat

Acryloid A-101	35
Vinyl Copolymer or Homopolymer	65
Paraplex G-62	Up to 3, if required
Cellulose Acetate Butyrate	4
Methyl Ethyl Ketone*	As required for proper solids or viscosity

* Blends of MEK and tetrahydrofuran may be required if higher molecular-weight vinyl or acrylic resins are used.

Acrylic Flexible Vinyl Substrate Topcoat

Acryloid A-101	15
Acryloid B-82	15
Vinyl Copolymer	70
Cellulose Acetate Butyrate	4
Methyl Ethyl Ketone	As required for proper solids or viscosity

Vinyl Automotive Top-Texturizer

Exon 470	11.00
Exon 450 or **Exon 497**	11.00
M-P-A (xylene)	1.65
FRN-505	1-3.00
FCR-1261	0.50
Syloid 308	1.50
Neo-Spectra AG Beads	1.50
MEK	25.00
MIBK	20.00
Toluene	30.00
Xylene	20.00

Procedure:

Make-up a 10% solution of the nitrile rubber in the automatic diluent using a Cowles mixer. If a high shear mixer is not available, the solution may be made-up on the rolls—heat will help dissolve the rubber.

Dissolve the vinyl resins in ketone solvent and remaining diluent (if **Exon 450** is formulated, heat may be required. High shear mixing should be employed). Add desired amount of nitrile rubber solution to the vinyl resin solution along with the remaining ingredients and ball mill. The use of MPA requires heat to properly disperse the thixotrope.

It may be desirous to make a predispersion of the thixotrope although this step is not required. Ball mill to an N.S. fineness of 6–8.

Acrylic Decorative Coatings

Formula	No. 1 *(For Polystyrene)*	No. 2 *(For ABS, Acrylic, Cellulosic)*
Acryloid B-66 (50%)	16.80	–
Acryloid B-72 (100%)	–	6.0
Sucrose Acetate Isobutyrate	3.36	–
Alcohol Soluble Butyrate	5.04	3.0
Half Second Butyrate	–	6.0
Pigment	7.20	5.0
Acetone	9.90	–
Isopropanol	25.40	–
Diacetone Alcohol	14.60	–
Toluene	14.70	40.0
Ektasolve EE Acetate	3.00	10.0
Denatured Ethanol	–	8.0
Ethyl Acetate	–	8.0
Isobutyl Acetate	–	14.0
Total Solids (%)	24.0	20.0
Visc. (cps)	30	45

Vinyl Plastisol Molding, Film Forming and Fabric Coating

Pony Mix

Geon 121	53.0
Dioctyl Phthalate	15.0
Dioctyl Adipate	6.0

Drapex 4.4	8.0
Ferro 5184 or **Mark GS****	3.0
Pigment, A or D-Series	15.0

Can be converted to an organosol by thinning with volatile diluents such as VM&P Naphtha or Xylene.

Fusion temperature 350 F (177 C).

A 6 mil thick fused film prepared from the above formulation with the D-Series pigment will show a moderate loss of color and brightness after 120 h exposure in an Atlas Twin-Arc Weather-Ometer.

The proper choice of light stabilizers is very important when the fluorescent pigments are used in a vinyl plastisol. The barium-cadmium type usually will either cause darkening of the fluorescent pigments during fusion, or very rapid darkening on exposure to direct sunlight. Much better light stability is obtained with the organic zinc type of stabilizers, of which the **Ferro 5184** or **Mark GS** are examples. Any other types of stabilizers used should be checked against the above formulation to be certain of improved or equivalent performance.

The following table gives the approximate minimum percentages of pigment for optimum color and light stability.

Thickness in Mils	Percent
3–5	15–30%
10	8–15%
20	4–8%
40 and over	1–4%

For most plastisols the highest practical working pigment content is 20%.

For organosols it is 30–35%.

In calendered vinyl film containing **Day-Glo** A or D-Series pigment, the use of **Mark 398** as a light stabilizer is recommended along with the usual epoxy heat stabilizer.

Vinyl Coating

Vehicle:

Diisobutyl Ketone
Vinyl Resin VMCC

High speed mix until all resin is in solution.

Pigmented Coating:

Vehicle	57.0
Pigment (D-Series)	43.0

High speed mix.

Fluorescent Acrylic Urethane Coating

A	**G-Cure 867-RWF-60**	16.4
	Exempt Solvent Blend*	8.9
	GT Series Pigment	35.6

Disperse on high-speed mixer to 50 C; add following under agitation:

G-Cure 867-RWF-60	17.8
Exempt Solvent Blend	12.3
Additive #11	0.1

Add to Part A before use to catalyze:

Desmodur N-75	8.9

---

* Special Rule 66 Exempt Solvent Blend

Rule 66 Exempt VM&P Naphtha	30.0
Ethyl Benzene	15.0
Cellosolve Acetate	55.0

Acrylic Urethane Clear Topcoat

A	**G-Cure 867-RWF-60**	56.3
	Additive #11	0.1
	Tinuvin 328	1.4
	Exempt Solvent Blend*	27.5

Add to Part A immediately before use to catalyze:

Desmodur N-75	14.7

---

* Special Rule 66 Exempt Solvent Blend

Rule 66 Exempt VM&P Naphtha	30.0
Ethyl Benzene	15.0
Cellosolve Acetate	55.0

General Purpose Phosphorescent

Phosphorescent Pigment		30.0
Acryloid B-72 (50%)		40.0
Toluene		10.0
Cellosolve Acetate		20.0
Total Solids (%)	50.0	

Note:

No grinding equipment is normally used for coatings of this type. To improve pigment suspension, aluminum stearate may be added.

Lacquers of this type should be applied over a solid-covering undercoat. Two coats are suggested for most work where the coating is to be exposed to heavy rainfall or very high humidity. A clear coat of **Acryloid B-72** reduced to 15-20% solids should be applied over the luminescent paint.

Epoxy-Coal Tar Coating

A	**Araldite 540 X-90**	281.3
	Coal Tar	380.1
	Magnesium Silicate	256.7
	Antisag Agent	10.9
	o-Dichlorobenzene	61.2
	n-Butanol	66.6
B	**Araldite Hardener 950**	20.2
	n-Butanol	20.2

High Build Epoxy Versamid Coating

		lb	gal
A	**Versamid 140**	162	20.0
	Thixatrol ST	6	0.7
	Rutile Titanium Dioxide	386	11.0
	Acicular #417	96	2.0
	Xylene	64	8.9
	Butanol	50	7.4

B	**Gen Epoxy 190**	376	38.4
	Xylene	68	9.4
	Beetle 216-8 (60%)	18	2.2

Procedure:

Charge all of Part A into pebble mill and grind 16-24 h. Part B is stirred in just prior to use.

PVC (%)	17.9
Vehicle NV (%)	73.7
Total NV (%)	84.8
NV by Vol. (%)	73.3

Note:

Thirty minutes induction period after mixing is recommended. Inadequate allowance for induction time results in lowered gloss.

High Build, Self-Priming Mastic Coating

		lb	gal
A	**Resypox 1660**	120.0	11.94
	Resypox 1577	120.0	12.57
	MIBK	72.0	10.78
	Butyl **Cellosolve**	8.0	1.07
	Toluene	80.0	11.04
	Red Lead (97%)	176.4	2.34
	Red Iron Oxide R-1599	50.0	1.16
	Barytes 89	190.1	5.30
	Mineral Filler 1718	190.1	8.91
	Bentone 18C	5.0	0.09
	MIBK	99.5	14.90
	Butyl **Cellosolve**	12.0	1.60
	Toluene	110.1	15.19
	Nuosperse 657	6.3	0.78
B	**Resymide 1115**	129.0	16.20
	Toluene	129.0	17.80

Procedure:

Prepare 60% NV solution of **Resypox 1660**. Charge pigments separately and blend thoroughly with good mechanical agitation. Add **Resypox** solution, **Nuosperse** mixture, and remaining solvents. Mix 30 min. Do not use high shear mixing, or excessive thixotropy will occur.

Do not mix any more material with curing agent than can be used in one working day.

Vinyl Acetate Silk Finish

Tioxide R-HD2		200.0
Speswhite		69.0
Natrosol 250 HHR (2% aq. sol'n.)		75.0
Calgon S (5% aq. sol'n.)		30.0
Sodium Benzoate (10% aq. sol'n.)		10.0
Potassium Hydroxide (50% aq. sol'n.)		2.0
Propylene Glycol		40.0
Butyl Dioxitol Acetate		10.0
Foamaster NXZ		3.0
Acticide MPM		0.4
Water		84.6
Vinamul 6705		462.0
Water	Premix	7.0
Tilcom AT21	Premix	7.0

PVC (%)	27
NV (%)	50
Visc. (poise—at 25 C, after shearing)	≈ 3
pH	7.8

Note:

The paint has good application properties and flow, ensuring a uniform semigloss or silk finish, and has good resistance to washing and scrubbing. It has a firm structure in the can, which may be reduced by decreasing the level of structuring aid, but this is likely to necessitate an increase in the amount of cellulosic colloid to maintain viscosity and flow.

Vinyl Acrylic Satin-Finish Board Coating

Mix together and disperse:

Tioxide R-HD2	200.0
SPS Clay	100.0
Natrosol 250 MR (2% aq. sol'n.)	90.0
Calgon S (5% aq. sol'n.)	30.0
Ammonia Sol'n. (≈ 0.880 sp. gr.)	1.2
Water	60.3

Then add:

Hercules 1512M	0.5
Acticide BG	1.0
Texanol	30.0
Vinacryl 4039	467.0
Polymul MS40	20.0

NV (%)	D 35
pH	≈ 7.5

Viscosity: adjust to give good coating properties under prevailing machine conditions.

Latex Board Coating

Formula No. 1

	Dry Weight
Clay (70%)	90.0
Titanium Dioxide	10.0
TSPP	0.1
α Protein (15%)	9.0
Latex (48%)	9.0
Water (to make 57% Total Solids)	

P/B Ratio	100:10
pH	10.1
Brookfield Visc. (cps)	3670

No. 2

	Dry Weight
Ultrawhite 90 Clay (60%)	90.0
Mississippi M60PI $CaCO_3$ (65%)	10.0
α Protein, LV (15%)	3.6
Latex (48%)	14.4
Water (to make 56.0% Total Solids)	
Calgon T (to disperse $CaCO_3$)	
P/B Ratio	100:18 (15.3%)

Latex Paper Coating

Formula No. 1

	Dry Weight
Astra-Glaze (70%)	100.00
Stayco-M (15%)	12.00
Latex (48%)	6.00
Parez-707 (70%)	0.77
Water (to make 51% solids)	
P/B Ratio	100:18
pH	7.1
Brookfield Visc. (cps)	2070

No. 2

	Dry Weight
Astra-Glaze (70%)	100.00
Stayco-M (15%)	7.00
Latex (48%)	7.00
Parez-707 (70%)	0.77
Water (to make 51% solids)	
P/B Ratio	100:14
pH	7.2
Brookfield Visc. (cps)	1110

Acrylic Cloth Coating

(For Books or Upholstery Fabric)

	Solids Basis
Acryloid B-72	30.0
Ethyl Cellulsoe	24.0
Monoplex DBS (dibutyl sebacate)	46.0

Note:

Reduce with toluene to 8% solids and apply by dipping. This finish gives excellent flexibility and is tack-free and nonyellowing.

Casein/Latex Offset Coating

Formula No. 1

	Dry Weight
Ultra-White 90 (70%)	90.0
Satin-White	10.0
Casein **FF-30** (ammonia cut–25%)	11.0
Latex	11.0
P/B Ratio	100:22
pH	10.0
Brookfield Visc. (cps)	17,500

No. 2

	Dry Weight
Ultra-White 90 (70%)	90.0
Satin-White (40%)	10.0
Casein **FF-30** (ammonia cut–25%)	4.4
Latex (48%)	17.6
P/B Ratio	100:22
pH	10.0
Brookfield Visc. (cps)	6480

Acrylic Gravure Ink

High speed mix:

Acryloid NAD-10	30.0
A or **AX** Pigment	50.0
Toluene	5.0

Thin with:

Heptane	15.0

Fluorescent Silk Screen Ink Vehicle

Aromatic Solvent **SC-100** (or super high flash naphtha)	30.0
EHEC High	6.0
Bentone Gel	4.0

Bentone 38	7
SC-100	90
Methanol (95%)	3

Made on high speed dissolver.

Pentalyn G	16.0
360 Mineral Spirits	34.0
Abitol 90X (this may be replaced in part by **Castung 403-UV**)	10.0

Procedure:

Dissolve resins in the aromatic solvent using a dissolver such as a Cowles and blend in other ingredients in order listed. Strain vehicle through nylon filter bags.

Fluorescent Screen Ink

(For General Purpose)

Above vehicle	60.0
A Series **Day-Glo** Pigment	35.0
Aromatic Solvent **SC-100**	5.0

Procedure:

Blend pigment into vehicle using high speed dissolver such as a Cowles and add solvent. Strain color through nylon filter bags. Final viscosity should be about 10,000 cps. Dilution to screening viscosity and wash can be done with mineral spirits.

Acrylic Ink for Polyvinylidene Film

CAP-504-0.2		9.7
SAIB		2.4
Acryloid B-72 (100%)		9.7
Pigment*		18.2
Isopropyl Alcohol (99%)		42.0
Water		18.0
Brookfield Visc. (cps–24 C)	1000	

* Concentration of pigment may vary with the color and degree of hiding desired.

Epoxy Thermoplastic Ink

(For Polyethylene and Polypropylene)

CAP-482-0.5		8.0
Uni-Rez 710		10.5
SAIB		2.0
Kodaflex DBP		5.2
Benzidene Yellow*		4.3
Tecsol C (anhydrous)		49.0
Ethyl Acetate (99%)		21.0
Brookfield Visc. (cps–24 C)	220	

* Concentration of pigment may vary with the color and degree of hiding desired.

Maleic Thermoplastic Ink

(For Polypropylene)

CAP-504-0.2		10.5
Arochem 404		10.5
SAIB		1.6
Chrome Yellow		17.5
Tecsol C (95%)		59.9
Brookfield Visc. (cps–24 C)	65	

Styrene-Alkyd Gravure Ink

High speed mix:

Cycopol 341-17	25.0
Antisettling Gel (20% Aluminum Stearate 222-2B in Special Naphtholite)	4.0
A or **A-Extra** Pigment	50.0
Special Naphtholite (K.B. Value 40)	10.0

Thin with:

Hexane	11.0

Urea Thermoplastic Ink

	Formula No. 1	No. 2
A CAP-482-0.5	10.8	–
CAP-504-0.2	–	12.5
Beckamine 21-511 (60%)	18.0	20.8
Tecsol C (anhydrous)	49.5	59.5
Ethyl Acetate (99%)	21.2	6.6
B *p*-Toluenesulfonic Acid (high para)	0.5	0.6

Procedure:

Add Part B to a solution of Part A. At room temperature, Formula GCRC-3 is stable approximately three months while FCRC-4 is stable approximately six months. These formulations may be pigmented to produce the desired color and degree of hiding. Both systems adhere to a wide variety of flexible substrates including treated polyethylene and polypropylene, shellac and nitrate coated aluminum foil, polyvinylidene chloride coated cellophane, uncoated cellophane, and paper.

No. 3

CAP-504-0.2	6.8
Beckamine 21-511 (60%)	11.3
SAIB	3.4
Kodaflex DBP	3.4

Pigment*	≈ 12.6
Tecsol C (anhydrous)	55.9
Ethyl Acetate (99%)	6.2
p-Toluenesulfonic Acid (high para)**	0.4
Brookfield Visc. (cps–24 C)	31

* Concentration may vary with the type and color of pigment.

** The acid catalyst should be added after the other components are in solution. This system is stable for approximately six months at room temperature.

Latex Tint Base for Cal Ink Colors

Formula No. 1

(Semigloss)

	lb	gal
Charge the following to the tank:		
Tamol 850 (30%)	9.0	0.91
Water	2.0	0.24
Propylene Glycol	60.0	6.96
Antifoamer 66	0.5	0.06
Add the following at low speed:		
Rutile Titanium Dioxide (high gloss)	225.0	6.57
Barytes	40.0	1.09

Grind the above in a high-speed mill (Cowles, 3800–4500 fpm for 10–15 min) and let down at lower speed with the following:

		lb	gal
Propylene Glycol		95.5	11.08
Rhoplex AC-490 (46.5%)		526.0	59.44
Acrysol G-110	Premix	10.3	1.22
Super Ad-It	Premix	1.0	0.12
Water	Premix	68.7	8.24
Butyl **Cellosolve**	Premix	24.5	3.12
Triton N-57		4.0	0.46
Triton X-207		4.0	0.49

PVC (%)	23.3
Total Solids by Vol. (%)	33.0
Total Solids by Wt. (%)	49.0
Visc.–Initial (KU)	68

No. 2

		lb	gal
Charge the following to the tank:			
Tamol 731 (25%)		5.5	0.59
Tamol 850 (30%)		4.6	0.46
Propylene Glycol		60.0	6.96
Add the following at low speed, 1900 fpm:			
Rutile Titanium Dioxide		225.0	6.57
Barytes		40.0	1.09
Increase speed to 5400 fpm and grind for 20 min; then add the following at low speed, 1900 fpm:			
Propylene Glycol	Premix	120.0	13.92
Preservative	Premix	1.0	0.12
Rhoplex AC-22 (44.5%)		440.0	49.72
Rhoplex AC-61 (46%)		110.0	12.40
Acrysol G-110 (22%)	Premix	10.3	1.23
Water	Premix	48.8	5.90
Triton N-57		6.0	0.69
Triton X-200 (28%)		6.0	0.67
Colloid 600		2.0	0.26

Total Solids (%)	48.6
PVC (%)	23.5
Visc.–Initial (KU)	70

No. 3

(Flat, Dripless)

	lb	gal
Charge the following to the tank:		
Water	43.0	5.16
Tamol 850 (30%)	6.0	0.60
KTPP	1.7	0.80
Hydroxyethyl Cellulose–2.5% **(QP-4400)**	159.0	19.08
Add the following at low speed, 1900 fpm:		
Hexylene Glycol } Premix	30.0	3.93
Ethylene Glycol } Premix	35.0	3.77
Super Ad-It } Premix	1.0	0.13
Antifoam 66 } Premix	0.5	0.06
Ti Pure R-901	160.0	4.82
1160 Silica	53.0	2.41
Icecap K	85.2	3.97
Zeolex 80	27.0	1.54
Increase speed to 4500 fpm and grind for 10 min, then let down at low speed, 1900 fpm:		
Rhoplex AC-22	277.5	31.36
Nopco NXZ	6.0	0.80
Isopar M	2.0	0.31
Veegum T (4% dispersion)	125.0	14.61
Hydroxyethyl Cellulose–2.5% **(QP-4400)**	30.0	3.60
Water	5.0	0.60
Triton GR-5	10.0	1.19
Triton X-114	5.0	0.57
Triton N-57	5.0	0.58

Adjust pH to 9.9 with ammonium hydroxide (28%).

PVC (%)	50.3
Total Solids (%)	45.4
Visc.–Initial (KU)	86

Alkyd Screen Process Color

Roller mill grind or high speed mix:

Duraplex DX 656	42.50
Aluminum Stearate-Kerosene Gel	8.90
Pigment, A or D-Series	45.00

Mix with:

Mineral Spirits	3.20
Cobalt Naphthenate (6% cobalt)	0.18
Manganese Naphthenate	0.12
Exkin No. 2	0.10

Thin to screen printing viscosity with mineral spirits.

Acrylic-Latex Swimming Pool Paint

Formula No. 1

(Blue, Semigloss)

	lb/100 gal
Paste:	
Titanox RANC	175
ASP 400	75
Phthalocyanine Blue	1
Bentone 11	4
Pliolite AC	60
Solvesso 100	110
Mineral Spirits	110
Let down:	
Pliolite AC	130
Dow 276-V2	18
Solvesso 100	122
Mineral Spirits	122

PVC (%)	27.1
Solids (%)	49.9
Visc. (KU)	80-82

Note:

For roller application reduce 1 qt solvent per 1 gal of paint.

No. 2

(Blue–Gloss)

	lb/100 gal
Paste:	
Titanox RANC	175
ASP 105	50
Phthalocyanine Blue	1
Bentone 11	4
Pliolite AC	40
Solvesso 100	80
Mineral Spirits	80
Let down:	
Pliolite AC	150
Dow 276-V2	18
Solvesso 100	156
Mineral Spirits	156
PVC (%)	24.4
Solids (%)	48.1
Visc. (KU)	78–80

Note:
For roller application reduce 1 qt solvent per 1 gal of paint.

Latex Swimming Pool Paint

Formula	No. 1	No. 2	No. 3
	(White)	*(Aqua)*	*(Blue)*
Aromatic High Flash Naphtha	210.0	210.0	210.0
Marbon 9200 MV	80.0	80.0	80.0
Mix to clear solution.			
Chlorinated Paraffin (50%)	80.0	80.0	80.0
Rutile Titanium Dioxide	225.0	160.0	160.0
Calcium Carbonate	175.0	215.0	215.0
Mica	70.0	70.0	70.0

Diatomaceous Silica	70.0	70.0	70.0
Suspension Additive*	4.0	4.0	4.0
Denatured Alcohol*	2.0	2.0	2.0

*Prewet

Mix to smooth paste and grind with 3-roll mill to fineness of 4+ N.S.

Mineral Spirits (KB40+)	210.0	210.0	210.0
Phthalocyanine Green in Oil	–	5.0	–
Phthalocyanine Blue in Oil	–	–	2.0

Procedure:

Add ingredients in order shown.

Total Solids (%)	62.6	61.3	61.1
PVC (%)	56.0	55.4	55.5
lb/gal	11.4	11.1	11.1
Visc. (KU)	85	85	85

Polyvinyl Chloride Swimming Pool Paint

	lb	gal
Xylene	361.9	49.85
Vinoflex MP 400 (medium)	185.4	18.00
Aroclor 1260	100.0	7.41
Titanox CLNC	175.0	5.08
Nytal 300	75.0	3.16
Barytes X-5-R	50.0	1.35
Bentone 38	10.0	0.44
Butyl Acetate	37.0	5.00
Mineral Spirits	65.0	10.00

Procedure:

Grind 16 h in a high density alumina jar mill.

PVC (%)	28.3
NV (%)	56.2
NV by Vol. (%)	35.3
lb/gal	10.6
Visc. (KU)	95 ± 5

Acrylic Latex Traffic Paint

Formula No. 1

(White, Beaded or Nonbeaded)

	lb/100 gal
Rutile Titanium Dioxide–Calcium ($CaSO_4$) (30/70) (MIL-T-15195A)	296
Rutile Titanium Dioxide (chalk resistant)	59
Magnesium Silicate	53
Diatomaceous Silica	83
Mica	59
Bentone 38	5
Soya Lecithin	8
Pliolite VTL	107
Hydrocarbon Resin	35
Chlorinated Paraffin (40%)	35
VM&P Naphtha (distillation range: 200–225 F)	357
Total Solids (%)	67.5
PVC (%)	51.5
Visc. (KU)	61

No. 2

(Yellow, Beaded or Nonbeaded)

	lb/100 gal
Medium Chrome Yellow (Type III-Federal Spec. TT-C-290)	259
Whiting	324
Magnesium Silicate	65
Bentone 38	4
Soya Lecithin	7
Pliolite VTL	105
Hydrocarbon Resin	35
Chlorinated Paraffin (40%)	35
VM&P Naphtha (distillation range: 200–225 F)	357
Total Solids (%)	70
PVC (%)	52.2
Visc. (KU)	62

No. 3

(White, Nonbeaded)

	lb/100 gal
Rutile Titanium Dioxide (NC)	189.5
Whiting	183.5
Diatomaceous Silica	94.5
Bentone 38	5.0
Soya Lecithin	7.5
Pliolite S-5A	118.5
Chlorinated Paraffin (40%)	57.0
Chlorinated Paraffin (70%)	62.0
Toluene	150.5
VM&P Naphtha	226.0
PVC (%)	42.8
Total Solids (%)	65.3
Visc. (KU)	72

No. 4

(White Beaded)

	lb/100 gal
Rutile Titanium Dioxide (NC)	140.5
Whiting	327.5
Asbestine	56.0
Pumice (No. 1)	56.0
Bentone 38	5.0
Soya Lecithin	9.5
Pliolite S-5A	105.5
Chlorinated Paraffin (40%)	51.0
Chlorinated Paraffin (70%)	55.0
Toluene	142.5
VM&P Naphtha	213.5
PVC (%)	52.0
Total Solids (%)	69.0
Visc. (KU)	80

No. 5

(Yellow, Nonbeaded)

	lb/100 gal
Chrome Yellow (medium)	245.0
Whiting	178.0
Diatomaceous Silica	92.5
Bentone 38	5.0
Soya Lecithin	7.5
Pliolite S-5A	115.5
Chlorinated Paraffin (40%)	55.5
Chlorinated Paraffin (70%)	60.0
Toluene	153.5
VM&P Naphtha	230.5
PVC (%)	42.8
Total Solids (%)	66.1
Visc. (KU)	70

No. 6

(Yellow, Beaded)

	lb/100 gal
Chrome Yellow (medium)	211.0
Rutile Titanium Dioxide (NC)	37.0
Whiting	265.5
Asbestine	62.0
Pumice (No. 1)	62.0
Bentone 38	5.0
Soya Lecithin	9.5
Pliolite S-5A	103.0
Chlorinated Paraffin (40%)	49.5
Chlorinated Paraffin (70%)	53.5
Toluene	144.5
VM&P Naphtha	217.0
PVC (%)	52.4
Total Solids (%)	70.0
Visc. (KU)	81

No. 7

(General Use)

CP Medium Chrome Yellow	302
Celite 281	35
Asbestine 3X	348
Pliolite S-5	70
Chlorowax 40	53
Piccodiene 2215	70
Lecithin	8
VM&P Naphtha	293
Toluene	64

PVC (%)	50
NV Solids (%)	71
Visc. (KU)	70-80

Procedure:

Charge all the **Pliolite S-5**, toluene, **Piccodiene** and 75 lb of VM&P Naphtha to a mixing tank and agitate until solution is complete. Charge this solution and all remaining ingredients to ball mill and grind.

	No. 8 *(Yellow)*	No. 9 *(White)*
Titanium Dioxide	–	15.0
Calcium Sulfate	–	15.0
CP Medium Chrome Yellow	25.2	–
Diatomaceous Silica	2.4	1.7
Talc	23.4	17.4
Pumice	4.2	2.5
Cumar Resin	5.8	7.5
Pliolite S-5	5.8	7.5
Chlorinated Paraffin	4.5	5.9
Petroleum Solvent (73 K.B.)	18.2	17.6
Methyl Ethyl Ketone	9.8	9.0
Lecithin	0.7	0.9

Alkyd-Chlorinated Rubber Traffic Line Paint

	No. 1	No. 2	No. 3	No. 4	No. 5	No. 6	No. 7	No. 8
Parlon S10	3.4	—	3.2	3.2	3.2	3.2	3.3	4.3
Parlon P10	—	3.4	—	—	—	—	—	—
Soybean-PE Alkyd (65%)	13.5	13.5	13.1	13.1	13.1	13.1	15.7	13.4
Titanox RCHT	45.2	45.2	—	—	—	—	7.5	44.0
Medium Chrome Yellow **X-2541**	—	—	14.0	14.0	14.0	23.4	13.0	—
Monarch Blue 12658	—	—	—	—	—	—	1.5	1.2
Asbestine 3X	5.6	5.6	14.7	14.7	33.7	5.3	17.7	5.6
Keystone Whiting	5.6	5.6	29.0	29.0	10.0	29.6	17.1	5.6
Soya Lecithin	0.7	0.7	—	—	—	—	0.6	0.7
MPA Dispersing Agent	—	—	0.4	0.4	0.4	0.4	—	—
Epichlorohydrin	0.1	0.1	0.1	0.1	0.1	0.1	0.1	0.1
Cobalt Naphthenate (6%)	0.1	0.1	—	0.1	—	—	0.1	0.1
Antiskinning Agent	0.1	0.1	0.1	0.1	0.1	0.1	0.1	0.1
Toluene	23.3	23.3	23.1	23.0	23.1	22.5	21.4	22.5
Mineral Spirits	2.4	2.4	2.3	2.3	2.3	2.3	1.9	2.4
PVC (%)	55.0	55.0	55.0	55.0	54.5	53.0	51.0	54.0
Total Solids by Wt. (%)	70.0	70.0	70.0	70.0	67.0	70.0	67.0	70.0
Visc. as applied (KU)	72	70	70	70	70	70	72	72

Alkyd Traffic Line Paint

(White)

	lb	gal
Rutile Titanium Calcium	625.00	23.00
Magnesium Silicate	74.00	3.30
Calcium Carbonate	72.00	3.20
Alkyd (60% NV)	314.00	38.50
Chlorinated Rubber	46.80	3.60
Benzene	71.00	9.70
Epichlorohydrin	0.90	0.10
Toluene	128.00	17.70
Post-4	8.00	0.92
Cobalt Naphthenate (6%)	1.50	0.18
Antiskinning Agent	1.25	0.16

PVC (%)	54.0
lb/gal	13.4

Chlorinated Rubber Traffic Paint

(Yellow)

	lb/100 gal
Chrome Yellow (medium)	230.0
Whiting	178.5
Talc	50.0
Diatomaceous Silica	80.0
Mica	60.0
Varnish (50% solids in toluene)*	247.0
Propylene Oxide	2.0
Chlorinated Paraffin	42.0
Chlorinated Rubber	86.0
Toluene	173.2
Aliphatic Thinner	76.8
Cobalt Naphthenate (6%)	1.3

Lead Naphthenate (24%)	2.5
Antiskinning Agent	1.2
Antisettling Agent	7.0 max.**

---

* Varnish Formula:

Castung 403 Z3	120
Piccodiene 2215	100
Toluene	220

Procedure—Varnish:

Dissolve resin in toluene and add the dehydrated castor oil with agitation.

** Note:

12 lb **M-P-A 60** (toluene) can be used.

lb/gal	7.76
Total NV (%)	50

Latex Traffic Line Paint

Formula No. 1

(Black)

	lb/100 gal
Styrene-Butadiene Resin	120.2
Chlorinated Biphenyl Resin (solid)	62.5
Chlorinated Biphenyl Resin (liquid)	57.7
Carbon Black	53.8
Aluminum Silicate Pigment	67.3
Diatomaceous Silica	33.7
Methyl Ethyl Ketone	463.5

Pigment by Wt. (%)	17.5-19.5
Visc. (KU) @ 77 F	60-70
lb/gal	8.5-8.8

No. 2

(Red)

	lb/100 gal
Toluidine Red	36.3
Molybdate Orange (dark)	19.5
Aluminum Silicate Pigment	148.9

Styrene-Butadiene Resin	135.4
Chlorinated Biphenyl Resin (solid)	70.4
Chlorinated Biphenyl Resin (liquid)	65.0
Methyl Ethyl Ketone	338.0
2-Nitropropane	112.7

Note:

M-P-A (toluene) is recommended for this specification to insure against caking and excessive settling of the pigment in the package as well as to provide for control of the viscosity. If **M-P-A 60** (toluene) is used, a level 1.65 times that of **M-P-A** (toluene) is recommended.

Pigment by Wt. (%)	21.0-23.0
Visc. (KU) at 77 F	60-70
lb/gal	9.1-9.4

No. 3

(White)

	lb	gal
Titanox RCHT-X	282.0	10.4
Titanox RANC	52.0	1.5
Magnesium Silicate (fibrous grade)	60.0	2.5
Celite 281	76.0	3.9
Mica	60.0	2.5
Organo-Montmorillonite	5.0	0.3
Pliolite VT (33.3% in VM&P Naphtha)	308.0	44.5
Chlorowax 40	32.5	3.4
Soya Lecithin	8.0	1.0
Velsicol X-37 (30.0% in VM&P Naphtha)	108.0	15.3
VM&P Naphtha	92.0	14.7

Pigment (%)	50.0
Vehicle (%)	50.0
Vehicle NV (%)	31.0
PVC (%)	53.5

No. 4

(White)

	lb	gal
Charge into mixer under agitation:		
Water	100.0	12.0
Cosan PMA-30	0.5	–
KTPP	1.0	0.1
Tamol 731	4.0	0.5
Igepal CO-630	4.0	0.4
Ti Pure R-901	200.0	5.9
Celite 281	50.0	2.6
Natrosol 250 HR (1½% sol'n.)	100.0	12.0
Fibrene C-400	100.0	4.2
Ethylene Glycol	25.0	2.6
Hexylene Glycol	15.0	2.0
Colloid 677 Defoamer	1.0	0.1
Disperse in high-speed disperser, then add slowly with agitation:		
Resyn 5005	400.0	44.0
Dibutyl Phthalate	15.0	1.7
Natrosol 250 HR (1½% sol'n.)	100.0	12.0
Imperial Black Dispersion **X-2488**	Trace	–

Note:

Protect **Natrosol** thickener solution with 0.08% **PMA-30** on total weight of solution, if stored prior to usage.

PVC (%)	36.5
CPVC (%)	39.5
lb/gal	11.1
Visc. (KU)	90-95
Total Solids (%)	51.5

	No. 5 *(Yellow)*	No. 6 *(White)*
Heptane (KB40+)	300.0	300.0
Marbon 1100T MV	120.0	120.0
Chlorinated Paraffin (50%)	62.0	62.0

Chlorinated Paraffin (70%)	19.0	19.0
Titanium-Calcium Pigment	50.0	460.0
Calcium Silicate (acicular)	190.0	125.0
Medium Chrome Yellow	190.0	–
Yellow Iron Oxide	25.0	–
Calcium Carbonate	175.0	–
Lecithin	2.0	2.0
Suspension Additive*	4.0	4.0
Denatured Alcohol*	2.0	2.0

*Prewet.

Grind in pebble mill to a fineness of 5+ N.S. Drain mill; then add:

Heptane (KB40+) to rinse mill	50.0	50.0

Procedure:

Add ingredients in the order shown.

Total Solids (%)	70.4	69.2
PVC (%)	49.8	50.6
lb/gal	11.7	11.2
Visc. (KU)	75	74

Polyvinyl Chloride Traffic Line Paint

	lb	gal
Lactol Spirits	120.0	16.28
Toluene	140.0	19.31
Acetone	90.0	13.53
Vinoflex MP 400 (medium)	175.0	16.94
Aroclor 1260	30.0	2.30
Resoflex F-296	5.0	0.63
Ketone Resin N	35.0	3.59
Titanox C-50	320.0	11.04
Aerosil COK 84	8.0	0.42
Asbestine 3X	75.0	3.24

Procedure:

Grind 16 h in a high density alumina jar mill.

PVC (%)	40.5
NV by Wt. (%)	64.8
NV by Vol. (%)	43.7
lb/gal	11.43
Visc. (KU)	89

Gloss White Enamel for Electrostatic Disc Spray

Sand mill (30 min)

Cymel 303	94.5
Tamol 731 (25%)	6.0
Butyl **Cellosolve**	12.0
Isopropanol	7.5
Water	130.0
DMAE (10% in water)	15.0
Rutile Titanium Dioxide	315.0

Let down:

Sand mill grind	250.1
Acrysol WS-68 (38%)	428.9
Water	68.6
Dipropylene Glycol	75.0
L-5310 (30% in butyl **Cellosolve**)	1.4
DMAE (10% in water)	132.0
PTSA* (10%)	4.0
n-Butanol	10.0

Total Solids (%)	35.0
P/B Ratio	40/60
pH	8.4
Visc. (s–#4 Ford Cup)	45

Note:

Use more DMAE to increase spray viscosity; use less DMAE to decrease spray viscosity.

---

* 10 PTSA/10 DMAE/80 Water

Latex "Hypalon" Paint for Rigid or Semirigid Substrates

Base Lacquer:

Hypalon 20	100.0
Staybelite Resin	2.5
Cellosolve Solvent	20.0
Xylene	205.0
Ionol	2.0

Dispersion:

Tri-Mal	40.0
Ti Pure R-900	80.0
Atomite Whiting	30.0
MBTS	1.0
Nuosperse 657	2.0
Cellosolve Solvent	30.0
Xylene	112.0
M-P-A 60 (xylene)	20.0

lb/gal	9.29
Total Solids by Wt. (%)	40
Total Solids by Vol. (%)	23-24
Visc. (cps)	3500
PVC (%)	31.8

Chapter IX

INDUSTRIAL PAINTS

These are coatings that are used on rubber, wood, paper, plastic, metal, glass, and any other type of manufactured object. Required characteristics are as varied as the market itself – from photosensitive materials to floral patterned wallpaper. They may be decorative or integrally involved in the product's function or both. Most, however, are designed for the large-scale, specific application techniques available in factories.

Acrylic Industrial White Enamel

Formula No. 1

(High Gloss, Air Dry)

	lb	gal
NeoCryl A-621	337.6	39.62
Nopcosant K	1.8	0.16
Nopco NXZ	1.3	0.18
Titanium Dioxide	212.9	6.09

Grind on high-speed impeller for 20 min @ 4100 fpm. Premix and add the following:

	lb	gal
NeoCryl A-621	455.2	53.43
Foamaster NS	0.9	0.12
BYK-301	1.6	0.20
DC-14	1.6	0.20

NV by Wt. (%)	53
PVC (%)	15
P/B Ratio	40:60

No. 2

(Med. Gloss, Air Dry)

	lb	gal
Grind Paste:		
Titanium Dioxide	199.6	5.99
Nopcosant K	3.0	0.28
Balab 3056A	3.0	0.41
Propylene Glycol	69.0	8.00
Ethylene Glycol	10.9	1.17
KTPP	0.3	0.01

Grind on high-speed disperser 20 min @ 5400 fpm.
Let down:

	lb	gal
Water	34.1	4.10
NeoCryl A-620	682.6	80.27
CO 610	3.2	0.36
Butyl **Cellosolve**	7.3	0.97
Balab 3056A	2.9	0.38
Ammonium Hydroxide (28%)	8.9	1.19

NV by Wt. (%)	43.5
PVC (%)	17
P/B Ratio	44:56
lb/gal	9.9
Visc. (cps)	3800
pH	9.8

	No. 3 *(Low Gloss, Air Dry)*		No. 4 *(Medium Gloss, Air Dry)*	
	lb	gal	lb	gal
Grind Paste:				
NeoCryl A-601	400.0	47.60	–	–
Titanium Dioxide	208.6	6.26	199.6	5.99
Tamol 731	3.4	0.41	3.0	0.28

Nopcosant K	6.8	7.83	–	–
Troy 999	–	–	3.0	0.41
Propylene Glycol	–	–	69.0	8.00
Ethylene Glycol	–	–	10.9	1.17
KTPP	–	–	0.3	0.01

Pebble mill overnight grind on high-speed disperser–20 min @ 5400 rpm.
Let down:

NeoCryl A-601	303.2	36.08	682.6	81.23
CO 610	–	–	3.2	0.36
Butyl **Cellosolve**	–	–	7.3	0.97
Balab 3056A	–	–	2.9	0.38
Ammonium Hydroxide (28%)	6.8	0.90	8.9	1.19

NV by Wt. (%)	44.3	43.5
PVC (%)	20	20
P/B Ratio	48:52	48:52
Visc. (cps)	2040	6400
pH	9.1	9.4

Acrylic Industrial Black Enamel

(Gloss, Air Dry)

	lb	gal
NeoCryl A-601	819.6	97.53
Tamol 731	5.2	0.63
Nopco NXZ	5.2	0.71
Neospectra MK11	13.1	0.79

Pebble mill overnight.

Ammonium Hydroxide (28%)	2.6	0.34

NV by Wt. (%)	33.3
PVC (%)	3
P/B Ratio	5:95
Visc. (cps)	500
pH	9.1

Acrylic Industrial Gloss Enamel

Formula	No. 1	No. 2	No. 3	No. 4
Cargill 7407	83.1	–	–	–
Beckasol 13-412	–	83.1	–	–
Chempol 10-0066*	–	–	83.1	83.1
Triethylamine	7.9	7.9	7.9	7.9
Water	49.4	49.4	49.4	49.4
Additive	–	–	–	4.5
Cobalt Drier (6%)	0.8	0.8	0.8	0.8
Manganese Drier (6%)	0.8	0.8	0.8	0.8
Zirconium Drier (6%)	0.8	0.8	0.8	0.8
Titanium Dioxide	240.3	240.3	240.3	240.3

Adjust pH to 8.0–8.5. Grind 10 min @ 3500 fpm to 7+ Hegman and let down with:

NeoCryl A-601	599.2	599.2	599.2	599.2
Water	41.5	41.5	41.5	41.5
NV by Wt. (%)	48.8	48.8	48.8	48.8
NV by Vol. (%)	36.0	36.0	36.0	36.0
PVC (%)	20	20	20	20
P/B Ratio	48:52	48:52	48:52	48:52
lb/gal	10.2	10.2	10.2	10.2
Visc. (cps)	104	286	80	80
pH	7.8	8.2	7.8	8.0

Alkyd Electrocoating Black Enamel

Formula	No. 1 *(Gloss)*	No. 2 *(Semi-gloss)*
Arolon 369-B-80	15.60	19.80
Butoxy-Ethanol	3.00	–
Triethylamine	1.00	2.01
Carbon Black	3.00	3.00
X5R	–	24.00

Oncor S25	–	6.00
Mineral Spirits (odorless)	0.50	–
Deionized Water	20.00	50.00
	Ball Mill	Pebble Mill

Bath Preparation

Arolon 369-B-80	109.10	75.40
Melamine Resin	9.70	6.70
Dimethylethanolamine	3.67	2.67
Guaiacol	0.48	0.34
Pigment Dispersion (as shown above)	43.10	104.81

Mix thoroughly adding water slowly until viscosity breaks.

Deionized Water	833.95	810.80
NV (%)	10.00	10.00
Pigment on Total Solids (%)	3.00	33.00
Suggested Bake	20 min 350 F	

Acrylic General-Purpose Baking Enamel

Formula	No. 1	No. 2	No. 3
	lb/100 gal	lb/100 gal	lb/100 gal
Rutile Titanium Dioxide	223.5	235.6	227.0
XC-4012 (48%)	174.4	140.0	226.6
Dimethylamino Ethanol	5.3	4.3	10.4
Demineralized Water	83.7	101.0	116.2
Cymel 303	56.0	–	–
Cymel 1123	–	87.9	–
BYK-P-104-S	2.8	2.9	2.8

Disperse on sand mill.

XC-4012 (48%)	285.2	281.5	128.6
Cymel 370 (88%)	–	–	126.0

Catalyst 4040	1.4	4.4	–
Dimethylamino Ethanol	8.7	8.5	5.8
Demineralized Water	186.4	174.1	189.3
Total NV by Wt. (%)	48.7	50.7	49.5
Total NV by Vol. (%)	36.9	38.6	37.2
PVC (%)	18.2	18.3	18.3

Acrylic White Baking Enamel

Formula No. 1

	lb	gal
Water	68.00	8.120
Dimethylamino Ethanol	1.40	0.170
Surfactol 365	6.40	0.730
Tamol 731	3.27	0.360
Troysan 174	0.30	–
Cymel 303	30.00	3.000
Tributyl Phosphate	0.35	0.045

Mix well, then add:

Titanium Dioxide	175.00	5.000

High speed grind 10 min. Let down, premix before adding grind.

Ubatol A-745	364.00	41.700
Water	237.32	28.340
Dimethylamino Ethanol	7.50	0.940
Butyl **Cellosolve**	41.43	5.470
Texanol	53.59	6.790
Tributyl Phosphate	0.35	0.045

Visc. (KU)	88
pH	9.0
Total Solids by Wt. (%)	36
Total Solids by Vol. (%)	23
lb/gal	9.23

No. 2

(White)

	lb	gal
Water	83.15	9.97
Amberlac 165	7.00	0.70
Tributyl Phosphate	1.00	0.13
Dimethylamino Ethanol	2.00	0.27
Mix well, then add:		
Titanium Dioxide	200.00	5.73
Mix well. Thin down with:		
Ubatol U-7403	668.00	75.50
Dimethylamino Ethanol (50% in water)	30.00	3.68
Butyl **Cellosolve**	58.00	7.38
Resimene X-745	25.00	2.50
Tributyl Phosphate	1.00	0.13

Total Solids (%)	49.4
Visc. (s—#4 Ford cup)	20-25
pH	9.0

White, High Gloss, Water-Reducible Baking Enamel

Formula	No. 1	No. 2	No. 3
	←	lb/100 gal	→
Unitane OR-650	410.3	416.7	423.7
XC-4011 (75%)	273.1	277.2	282.1
Dimethylamino Ethanol	21.2	21.5	25.8
Disperse on 3-roll mill and add:			
XC-4011 (75%)	273.5	206.7	142.4
Cymel 303 (100%)	102.8	–	–
Cymel 1123 (100%)	–	155.6	–
Cymel 370 (88%)	–	–	241.2

Butanol	44.2	49.5	57.0
Dimethylamino Ethanol	21.9	15.7	13.0
Catalyst 4040	2.6	7.8	–
Total NV by Wt. (%)	80	80	80
Total NV by Vol. (%)	65	65	65
PVC (%)	18.5	18.4	18.9

Acrylic-Epoxy White Baking Enamel

	lb	gal
Roller mill grind:		
Rutile Titanium Dioxide	403.0	11.58
Acryloid AT-75 (50% solids)	269.0	32.00
Mix with:		
Acryloid AT-75 (50% solids)	159.0	18.90
Araldite 6060 (50% **Cellosolve** acetate)	232.0	25.70
Solvesso 150	65.5	8.88
Cellosolve Acetate	21.8	2.83
Raybo 3	0.5	0.07
lb/gal	11.5	
Total Solids (%)	63.7	
Visc. (s–#4 Ford Cup)	85	
Total Vehicle (%)	45	

White High Gloss Baking Enamels

Formula	No. 1 *(Water-Based)*	No. 2 *(Alkyd-Based)*	No. 3 *(Acrylic-Based)*	No. 4 *(Polyester-Based)*
Rutile Titanium Dioxide	169.0	321.4	288.5	338.2
XC-4011 (75%)	112.5	–	–	–
Rezyl 387-5 (55%)	–	194.8	–	–
Acryloid AT-56 (50%)	–	–	192.3	–

Cyplex 1476-8 (65%)	–	–	–	173.4
Dimethylamino Ethanol	8.7	–	–	–
Xylol	–	19.5	–	–
Solvesso 150	–	–	–	14.5
Disperse and add:				
XC-4011 (75%)	56.8	–	–	–
Rezyl 387-5 (55%)	–	320.9	–	–
Acryloid AT-56 (50%)	–	–	322.0	–
Cyplex 1476-8 (65%)	–	–	–	265.4
Cymel 380 (80%)	105.8	88.7	80.4	112.5
Triethylamine	–	3.5	–	–
Dimethylamino Ethanol	3.1	–	–	–
Isophorone	–	–	–	38.4
Ethylene Glycol Monoethyl Ether	–	–	–	7.7
Methyl Isobutyl Ketone	–	–	40.7	–
Xylol	–	77.6	53.2	–
Butanol	13.3	40.3	39.5	15.9
Demineralized Water	501.7	–	–	–
Solvesso 150	–	–	–	131.6
Total NV by Wt. (%)	39.2	63.7	60.0	65.0
Total NV by Vol. (%)	26.8	45.8	42.4	48.1
PVC (%)	18.9	20.0	20.4	21.1
Visc. (s–#4 Ford Cup)	70	60	36	88

Alkyd White Bake Enamel

Formula No. 1

	lb	gal
Arolon X322 (at 50% triethylamine neutralization)	100.2	11.40
Ti Pure R-902HG	199.5	6.00
Water	105.5	12.67
Pebble mill for 16 h.		
Arolon X322 (at 50% triethylamine neutralization)	260.4	29.62
Resimene X-720 (80%)	25.1	2.51

Propasol B	3.5	0.48
Tri-n-Butyl Phosphate	8.1	1.00
Water	302.1	36.32

Total NV (%)	39.8
P/B Ratio	1.0:1.0
Percent Melamine–Solids Basis	10.0
Visc. (s–#4 Ford Cup)	50-60
pH (as is)	7.5-8.0
Suggested Bake	30 min/250 F

No. 2

	lb	gal
Ti Pure R-902HG	226.0	6.78
Arolon 580-W-42	113.0	12.71
Cellosolve	45.0	5.81
Water	22.0	2.64
Mineral Spirits (odorless)	5.4	0.86

Pebble mill 18-24 h.

Arolon 580-W-42	526.0	59.17
Resimene X-720 (80% NV)	41.0	4.10
n-Butanol	30.0	4.44
Water	28.6	3.43
Sodium Nitrite	1.0	0.06

P/B Ratio by Wt.	0.75:1.00
Total NV (%)	52.8
Visc. (s–#4 Ford Cup)	40--60
pH (adjust with triethylamine if low)	7.0-7.5

No. 3

	lb	gal
Arolon 376-WA8-50	105.44	12.12
Ti Pure R-902HG	210.58	6.32
Water	92.21	11.07

Pebble mill 18–24 h.

Arolon 376-WA8-50	274.05	31.50
Resimene X-720	26.40	2.64
Propanol B	3.73	0.51
Tri-n-Butyl Phosphate	8.44	1.04
Water	289.88	34.80

P/B Ratio	1:1
Paint Solids (%)	41.7
Visc. (s—#4 Ford Cup)	50-60

No. 4

(Semigloss)

	lb	gal
Arolon 557 (50% NV)	117.78	14.19
Micronized Talc	47.25	2.10
Colloidal Silica	1.75	0.10
Ti Pure R-902	187.81	5.64
Deionized Water	129.61	15.56

Pebble mill 18–24 h.

Arolon 557 (50% NV)	235.80	28.41
Cymel CLA-1125-85-37	34.39	3.62
Uformite MM-83	36.67	3.86
Isopropanol	15.52	2.37
Deionized Water	201.17	24.15

P/B Ratio	1:1
Paint NV (%)	46.9
Visc. (s—#4 Ford Cup)	50-60

No. 5

(Acrylic-Polyester Modified)

	lb	gal
Arolon 557 (50%)	106.85	12.72
Ti Pure R-902	236.91	7.11
Water	122.87	14.75

Pebble mill 18–24 h.

Arolon 465-WA8-70	226.80	25.20
Cymel 303	54.40	5.44
L-5310 (10%)	5.34	0.71
Water	283.80	34.07
P/B Ratio	0.89	
Total Solids (%)	48.6	
Visc. (s—#4 Ford Cup)	60–70	
pH @ 25 C	8.2–8.6	

No. 6

(Urea-Formaldehyde Modified)

	lb	gal
Roller mill grind:		
Titanium Dioxide	315	9.7
Duraplex ND-78 (60% solids)	210	24.6
Mix with:		
Duraplex ND-78 (60% solids)	105	12.3
Uformite F-240 (60% solids)	105	12.4
Uformite F-200E (50% solids)	126	14.9
Xylene	189	26.1
lb/gal	10.5	
Total Solids (%)	60	
P/B Ratio	1:1	

Note:

The enamel should be reduced to 55% solids with xylene for spray application and should be baked for 30 min at 300 F.

No. 7

	lb	gal
Titanox RA-50	218.0	6.22
Barytes X-10R	218.0	5.99
M-P-A (xylene)	0 or 7.5	0 or 1.03

Beckosol 1307-50HV	500.0	61.80
Uformite 200E	45.5	5.41
Solvesso 100	114.5	19.60

Procedure:

Ball mill at approximately 110 F.

PVC (%)	29.7
lb/gal	11.0
Visc. after 1 week (KU)	83

No. 8

(Gloss)

	lb	gal
Ti Pure R900	180.0	5.30
M-P-A 60 (xylene)	3.5	0.48
Butyl **Cellosolve**	10.1	1.35
Stepan Alkyd 4209	72.0	8.37

Three roll mill grind, 3 passes.

Stepan Alkyd 4209	309.0	35.91
Resimene 731	59.6	6.21
SF-69 (2% in **Cellosolve** acetate)	2.5	0.31
Chevron #3	25.9	3.50
Ethyl Benzene (99%)	38.0	5.25
Butyl **Cellosolve**	272.5	36.32

Total Solids by Wt. (%)	43.1
Visc. (s—#4 Ford at 25 C)	84.6

Oil-Free Alkyd Baking Enamel

Titanium Dioxide **RA-45**	315
M-P-A 60 (xylene)	6
Beckosol 93-642	490
Super Beckamine 27-580	105
Toluol	160

Methyl Ethyl Ketone	42
Butyl **Cellosolve**	36
Xylol	56
Silicone 200 Fluid (1% sol'n.)	3

Visc.–Initial (s–#4 Ford Cup)	50–60
lb/gal	12.2
Baking Time	15 s @ 300 F

Alkyd Gloss Baking Enamel

Fomula No. 1

	lb	gal
Varkyd 606-60AL	164	20.0
Super VM&P Naphtha	96	15.0
n-Butanol	34	5.0
M-P-A 60 (xylene)	6	0.7
Ti Pure R-900	275	7.9

Sand or pebble mill:

	lb	gal
Varkyd 606-60AL	213	26.0
Super VM&P Naphtha	45	7.0
Butyl **Cellosolve**	15	2.0
Melamine VN-812	155	18.0

PVC (%)	20.2
Total Vehicle NV by Wt. (%)	44.0
Total NV by Wt. (%)	59.3
lb/gal	9.88
Visc. (KU)	60–65
Bake Schedule	20 s @ 300 F

No. 2

	lb	gal
Varkyd 1577-45 VMJ	266	35.0
M-P-A 60 (xylene)	8	0.9
Butyl **Cellosolve**	30	4.0

Nuosperse 657	4	0.5
Solvent **102J**	14	2.0
Ti Pure R-900	275	7.9

Sand grind or pebble mill:

Varkyd 1577-45 VMJ	350	46.0
Melamine VN-812	56	6.5

PVC (%)	20.5
Total Vehicle NV by Wt. (%)	42.2
Total NV by Wt. (%)	55.8
lb/gal	9.77
Bake Schedule	20 s @ 300 F

Alkyd Short Oil Baking Enamel

Paste:

Unitane OR-540	625
Alkyd Resin Formulation Short Oil	375

"A" grind on 3 roll mill
Pigment/nonvolatile vehicle–3:1

Enamel:

Paste (above)	518
Alkyd Resin Formulation Short Oil	273
Cymel 248-8	115
Butanol	60
Xylene	34

lb/gal	11.7
Visc. (s–#4 Ford Cup at 25 C)	60

Alkyd Medium Oil Baking Enamel

Acintol Tall Oil Fatty Acid	714
Phthalic Anhydride	444
Pentaerythritol (tech.)	282
Ethylene Glycol	39

Procedure:

Charge all materials to a 2-ℓ reactor fitted with thermometer, inert gas and stirrer inlets. Adjust inert gas flow to 0.15 cfm/gal. Heat to 240 C and hold for an acid number of 4–8. Cool and reduce to 50% solids with mineral spirits. Approximate time of cool: 6–7 h.

Alkyd Blue Metallic Baking Enamel

Formula No. 1

		lb	gal
A	**Alcoa 726**	3.64	0.29
	Butyl **Cellosolve**	11.79	1.57
B	**Arolon 557** ((50% NV—dimethylethanolamine neutralization)	323.40	38.50
	Resimene X-720 (80% NV)	87.40	8.74
	Tributyl Phosphate	9.74	1.20
	Water	399.67	47.98
C	**Arolon 557**	14.83	1.72

Procedure:

Add Part A to Part B and then Part C.

Visc. (s—#4 Ford Cup) 30-40

No. 2

	lb	gal
Phthalocyanine Blue BT-417-D	8.27	0.62
Arolon 557 (50% NV)	41.32	4.92
Water	50.41	6.05

Procedure:

Steel ball mill 16–18 h.

Epoxy Baking Enamel and Primer

Formula	No. 1	No. 2	No. 3	No. 4	No. 5	No. 6	No. 7	No. 8
	(Yellow)		*(Green)*		*(Blue)*		*(Black Primer)*	
TMA Resin 408 (80% NVM)	255	255	255	255	255	255	255	255
Triethylamine	18	18	18	18	18	18	18	18
MM-83 (80% NVM)	45	–	45	–	45	–	45	–
Cymel 301	–	36	–	36	–	36	–	36
Water	444	453	434	443	434	443	476	485
Mn Naphthenate	2	2	2	2	2	2	2	2
Cadmium Yellow X-2826	80	80	–	–	–	–	–	–
Sherwood Green A-4436	–	–	40	40	–	–	–	–
Monastral Blue BT	–	–	–	–	40	40	–	–
Ti Pure R-900	16	16	8	8	8	8	–	–
Raven 150	–	–	–	–	–	–	6	6
Vera Blanc	–	–	–	–	–	–	94	94
pH–Initial	← 7.5-8.0 →							
Visc. (s–#4 Ford Cup)	← 35-45 →							
P/B Ratio	← 0.4:1.0 →		← 0.2:1.0 →		← 0.2:1.0 →		←0.42:1.0 →	

Alkyd High Gloss Yellow Baking Enamel

Medium Chrome Yellow **DCC1000**	300
M-P-A 60 (xylene)	6
Super Beckamine 27-580	105
Beckasol 93-642	490
Toluol	180
Methyl Isobutyl Ketone	62
Butyl **Cellosolve**	36
Xylol	50
Silicone Additive #1	1

Visc.–Initial (s–#4 Ford Cup)	50
lb/gal	12.30
Baking Schedule	15 min @ 300 F

Epoxy White Baking Enamel

(Economical)

Rutile Titanium Dioxide	24.00
Soya Alkyd Resin*	56.11
Uformite F-200 E	6.34
Amsco Solvent B	9.14
VM&P Naphtha	3.41
Tint with paste of fast Bordeaux Violet **B-11**	

* Phthalic anhydride, 36% solids, 50% solvent, xylol. Visc. T–V.

Procedure:

Melt all colors and powders together in a pebble mill. Then mix the liquids and the oils together–if necessary–heat the mixture until dissolved. Mix with the ground color and powders.

Polyester White High Gloss Industrial Baking Enamel

Formula	No. 1	No. 2	No. 3	No. 4
		lb/100 gal		
Unitane	219.4	219.9	219.8	297.7
Cyplex 1473-5 (65%)	59.6	59.7	59.7	152.6
Ethylene Glycol Monoethyl Ether Acetate	22.9	–	–	–
Solvesso 100	46.6	46.7	46.7	–
Xylol	–	–	–	16.4
Disperse and add (mill):	sand	sand	sand	3-roll
Cyplex 147306 (65%)	288.5	248.8	236.8	230.0
Cymel 301 or **303**	39.9	–	–	–
½″ Cellulose Acetate-Butyrate (20% in MIBK)	39.9	40.1	–	–
Catalyst 1010	2.7	–	–	–
Cross-Linking Agent 370 (88%)	–	75.9	–	–
Cymel 248-8 (55%)	–	–	158.1	–
Beetle XB-1032 (50%)	–	–	–	165.8
Ethylene Glycol Monoethyl Ether Acetate	0.9	0.9	0.9	–
Solvesso 100	24.6	24.6	24.6	–
Methyl Isobutyl Ketone	15.5	15.5	40.4	21.0
Butanol	45.3	38.5	0.6	46.8
NV by Wt. (%)	51.0	51.0	51.0	60.0
Visc. (s–#4 Ford Cup)	17	16	15	33
PVC (%)	19.6	19.7	19.7	20.9

Industrial Interior Enamel

Formula No. 1

(Green)

	lb	gal
Imperial X-1134	98.96	2.33
Arolon 969 (30%–ammonia neutralization)	197.93	22.96
Water	45.68	5.50

Pebble mill to 7½ Hegman.

Arolon 969 (30%—ammonia neutralization)	401.77	53.58
Manganese Naphthenate (6%)	0.65	0.08
Cobalt Naphthenate (6%)	0.65	0.08
n-Butanol	30.45	4.51
Water	91.20	10.96

NV by Wt. (%)	32.1
P/B by Wt.	0.5:1.0
Visc. (s—#4 Ford Cup)	50-70
pH—Initial	8.0-8.5

No. 2

(Black)

	lb	gal
Peerless 155	27.50	1.91
Arolon 465-WA8-70	118.35	13.15
Tamol 850	1.48	0.15
Distilled Water	16.57	2.06

Steel ball mill 48 h.

Arolon 465-WA8-70	328.32	36.48
Resimene X-720 (80%)	97.60	9.76
Propasol B	62.95	8.60
L-5310 (10%)	3.50	0.42
Distilled Water	228.82	27.47

NV by Wt. (%)	40.0
Visc. (s—#4 Ford Cup)	40-60

White Epoxy Enamel—Solventless, Thick Film

	lb	gal
Epon 815	534	56.3
Rutile Titanium Dioxide (nonchalking)	81	2.4
Sparmite	194	5.5

Celite 165-S	49	2.6
Thixatrol ST	10	1.2
Silicone Resin SR-82	8	0.9
Phenol Sol'n.–50% Wt. in **Neosol** Proprietary Solvent	32	4.4
Curing Agent Component:		
Epon Curing Agent H-3	215	26.6

Application:

For use over rigid substrates where appearance, as well as protection, is important, particularly for tank linings, masonry coatings, and marine coatings.

Epoxy Maintenance Coating

Formula No. 1

		lb	gal
A	**Epon 1001**	171.76	17.35
	Cellosolve	43.00	5.55
	OR-580	270.67	7.85
	Bentone 27	2.98	0.21
	Methyl Alcohol	6.40	0.97
	ASP-400	78.87	3.67
	Toluene	39.89	5.51
	Methyl Ethyl Ketone	45.23	6.73
	Methyl Butyl Ketone	45.43	6.73
	n-Butyl Alcohol	35.11	4.53
	Cellosolve	107.64	15.90
B	**Uni-Rez 2115C70**	110.33	13.69
	Cellosolve	72.62	9.37
	Xylene	14.06	1.94

Application:

This coating offers excellent long term protection to wood, concrete, aluminum, and primed steel surfaces. Recommended application is by spray, brush, or roller.

No. 2

		lb	gal
A	**Epon 1001**	172.26	17.40
	Cellosolve	43.01	5.55
	OR-580	271.36	7.87
	Bentone 27	3.00	0.21
	Methyl Alcohol	6.47	0.98
	ASP-400	79.08	3.68
	Toluene	31.42	4.34
	Methyl Ethyl Ketone	45.36	6.75
	Methyl Butyl Ketone	45.56	6.75
	n-Butyl Alcohol	107.91	15.94
	Cellosolve	42.86	5.53
B	**Uni-Rez 2125**	57.36	7.17
	n-Propyl Alcohol	98.49	14.70
	Toluene	22.66	3.13

Application:

This coating offers excellent long term protection to wood, concrete, aluminum, and primed steel surfaces. Recommended application is by spray only because of the fast drying time.

No. 3

		lb	gal
A	**Epon 1001**	166.82	16.85
	Cellosolve	41.70	5.38
	OR-580	262.74	7.62
	Bentone 27	2.84	0.20
	Methyl Alcohol	6.27	0.95
	ASP-400	76.72	3.57
	Toluene	47.71	6.59
	Methyl Ethyl Ketone	43.95	6.54
	Methyl Butyl Ketone	44.15	6.54
	n-Butyl Alcohol	104.46	15.43
	Cellosolve	41.31	5.33

B **Uni-Rez 2115P80**	93.79	11.95
n-Propyl Alcohol	76.58	11.43
Toluene	11.73	1.62

Application:
This coating offers excellent long term protection to wood, concrete, aluminum, and primed steel surfaces. Recommended application is by spray, brush, roller.

No. 4

(High Solids, Gray)

	lb	gal
A **Epon 828**	323.59	33.36
Epoxide 7	36.02	4.74
OR-580	435.14	12.62
Regal 330	14.55	0.97
Bentone 27	35.78	2.52
Methanol	14.45	2.19
Asbestine 325	89.50	3.77
Novacite 325	53.88	2.45
Beetle 216-8	18.02	2.12
Ethylene Glycol Monoethyl Ether	14.96	1.93
B **Uni-Rez 2341**	143.91	17.55
Toluene	17.96	2.48
n-Butanol	21.66	3.20
Ethylene Glycol Monoethyl Ether	78.28	10.10

Application:
This coating cures at room temperature overnight and develops good resistance to water, salts, alkalies, and dilute acids. Recommended application is by brush, roller, or spray to primed steel or concrete at a thickness of up to 0.020 in. per coat.

No. 5

		lb	gal
A	**Epon 1001**	145.53	14.70
	Cellosolve	36.42	4.70
	OR-580	356.18	10.33
	Bentone 27	2.41	0.17
	Methyl Alcohol	5.41	0.82
	ASP-400	66.83	3.11
	Toluene	33.81	4.67
	Methyl Ethyl Ketone	38.30	5.70
	Methyl Butyl Ketone	38.48	5.70
	n-Butyl Alcohol	91.19	13.47
	Cellosolve	25.58	3.30
B	**Uni-Rez 2100P75**	194.04	24.99
	n-Propyl Alcohol	35.64	5.32
	Toluene	11.95	1.65
	Cellosolve	10.62	1.37

Application:

This coating offers excellent long term protection to wood, concrete, aluminum, and primed steel surfaces. Recommended application is by spray, brush, or roller.

No. 6

(Polyamide-Modified)

Resin:

Titanium Dioxide			390.60
Bentone 27			4.26
Methanol			2.13
Epoxy Resin	156.60	Precut	
Xylene	52.20	Precut	
Cellosolve	52.20	Precut	

Two passes on three roll mill to 7+ N.S. fineness. Let down with:

Epoxy Resin	148.86	Precut	248.10
Xylene	49.62		
Cellosolve	49.62		
Methyl Isobutyl Ketone			9.31
Xylene			56.40
Cellosolve			28.20

Visc. (s–No. 4 Zahn): 34
Wt./gal (lb): 12.55

Hardener:

Uni-Rez 2100	Same as **Uni-Rez 2401** which may be used	304.0
Cellosolve		20.2
Xylene		182.0
Butyl Alcohol		25.3
Cellosolve		50.6

Visc. (s–No. 4 Zahn): 34
Wt./gal (lb): 7.59

Procedure:

Blend resin and hardener together, let stand for a 30 min induction period, filter before using. May be applied by brush, roller, or spray. May be reduced as necessary.

Application:

For application to primed steel, aluminum, wood, and concrete by spray, brush, or roller.

Spray Reducer:

Xylene	80
Cellosolve	20

Epoxy-Versamid Flat Coating

(Photochemically Nonreactive Solvent System)

	lb	gal
Epoxy Component:		
Genepoxy 451CS 60	356	40.0

Coreactant Component:

Diad 30	127	16.7
Versamid 230TP75	63	8.0
Ti Pure R-902	570	17.2
Asbestine 325	580	24.9
Celite 281	290	15.1
M-P-A 60 (xylene)	18	2.4
Solvesso 150	24	3.3
Cellosolve	272	35.0
Butyl **Cellosolve**	60	8.0
Mineral Spirits	190	29.4

Total Solids by Wt. (%)	70.8
Total Solids by Vol. (%)	48.7
Visc. (KU at 77 F)	78

Epoxy-Versamid Gel Coat

		lb/100 gal
A	**ERL-2795**	573.35
	Ti Pure R-610	46.09
	Zinc Oxide (**XX-50**)	22.79
	Antimony Oxide (**Red Star**)	11.57
	Alpine Talc #111	28.53
	Thixatrol ST	27.16
B	**Versamid 140**	252.72
	DMP-30	7.54
	Ti Pure R-610	31.08
	Cobalt Blue Pigment	3.38
	Butanol	8.24

Procedure:

All ingredients of Part A are blended in one-third the resin in disperser type equipment to a Hegman 5H grind gauge reading, after which the remainder of resin is added. Repeat operation for Part B. Part A and Part B are combined at use.

Application:

Thick film coating for general applications, appliance finishes, protective and maintenance finishes for metal and wood. Brush, trowel, spray.

Thick Film-Corrosion Resistant Epoxy Coating

Formula No. 1
(Ambient Cure)

A	**Araldite 6005**	148.9
	Titanox RA-50	22.4
	Thixatrol ST or **Thixcin R**	7.5
B	**Araldite DP-131-H**	89.2
	Dibutyl Phthalate	10.7
C	**Araldite DP-131-A**	1.5

Procedure:

Thixatrol ST is suggested if Part A is to be mixed on disperser type equipment with an approximate shearing temperature of 150 F. For a three roll mill, **Thixcin R** is the preferred thixotrope.

Application:

Recommended where a hard, tough, high-build, corrosion resistant coating is required. Brush, trowel, short nap roller. The use of **Thixatrol ST** or **Thixcin R** results in a pronounced improvement in the handling and workability of this compound.

No. 2
(Polyamide Cure)

A	**Versamid 140**	150.0
	Titanox RA-50	360.0
	Eagle Picher #417	90.0
	Xylene	57.0
	Butanol	45.0
	M-P-A 60 (xylene) or	14.9
	Thixatrol ST	5.4
B	**Gen Epoxy 185**	350.0
	Xylene	66.0

Procedure:

In this formulation, **Thixatrol ST** is generally preferred for a ball mill grinding operation and **M-P-A 60** (xylene) for a disperser type incorporation.

Triamino Ester Chrome Yellow Maintenance Enamel

	lb	gal
Chemacoil TA303 (80%)	300.0	39.22
Medium Chrome Yellow	350.0	7.49
Methyl Ethyl Ketoxime	2.0	0.20
Thixcin 25C	1.0	0.15
Grind 7–8 N.S. and add:		
Chemacoil TA303 (80%)	175.0	22.87
Cobalt Naphthenate (6%)	2.5	0.31
Lead Naphthenate (24%)	8.0	0.80
Zinc Naphthenate (8%)	4.0	0.40
Mineral Spirits	186.0	28.56

lb/gal	10.29
PVC (%)	13.7
Total Solids by Wt. (%)	70.9
Total Solids by Vol. (%)	54.7
Visc.–Initial (KU)	69

Styrene-Acrylic Maintenance Paint

Tioxide R-CR2	250.0
Microcal ET	10.0
D. Powder (China Clay)	145.0
Celacose CP10 (2% aq. sol'n.)	160.0
Vinapol 1640 (diluted 1:6 with water)	30.0
Water	68.6
Ammonia (sp. gr. 0.880–0.885)	2.0
Dalpad A	15.0
Foamaster NXZ	2.0

Acticide MPM	0.4
Vinacryl 7172	297.0
Empilan 2020 (10% aq. sol'n.)	20.0
PVC (%)	50
NV (%)	55

APPENDIX

*p*H Values

Acids	*p*H Value	Bases	*p*H Value
Hydrochloric Acid	1.0	Sodium Bicarbonate	8.4
Sulfuric Acid	1.2	Borax	9.2
Phosphoric Acid	1.5	Ammonia	11.1
Sulfuric Acid	1.5	Sodium Carbonate	11.6
Acetic Acid	2.9	Trisodium Phosphate	12.0
Alum	3.2	Sodium Metasilicate	12.2
Carbonic Acid	3.8	Lime, Saturated	12.3
Boric Acid	5.2	Sodium Hydroxide	13.0

*p*H Ranges of Common Indicators

	Useful *p*H Range
Thymol Blue	1.2 – 2.8
Bromphenol Green	2.8 – 4.6
Methyl Orange	3.1 – 4.4
Bromcresol Green	4.0 – 5.6
Methyl Red	4.4 – 6.0
Propyl Red	4.8 – 6.4
Bromcresol Purple	5.2 – 6.8
Brom Thymol Blue	6.0 – 7.6
Phenol Red	6.8 – 8.4
Litmus	7.2 – 8.8
Cresol Red	7.2 – 8.8
Cresolphthalein	8.2 – 9.8
Phenolphthalein	8.6 – 10.2
Nitro Yellow	10.0 – 11.6
Alizarin Yellow R	10.1 – 12.1
Sulfo Orange	11.2 – 12.6

International Atomic Weights

	Symbol	*Atomic Number*	*Atomic Weight*		Symbol	*Atomic Number*	*Atomic Weight*
Actinium	Ac	89	227	Neodymium	ND	60	144.27
Aluminum	Al	13	26.98	Neptunium	Np	93	237.00
Americium	Am	95	243	Neon	Ne	10	20.183
Antimony	Sb	51	121.76	Nickel	Ni	28	58.71
Argon	A	18	39.944	Niobium	Nb	41	92.91
Arsenic	As	33	74.91	Nitrogen	N	7	14.008
Astatine	At	85	210	Osmium	Os	76	190.20
Barium	Ba	56	137.36	Oxygen	O	8	16
Berkelium	Bk	97	249	Palladium	Pd	46	106.4
Beryllium	Be	4	9.013	Phosphorus	P	15	30.975
Bismuth	Bi	83	209.00	Platinum	Pt	78	195.09
Boron	B	5	10.82	Plutonium	Pu	94	242.00
Bromine	Br	35	79.916	Potassium	K	19	39.100
Cadmium	Cd	48	112.41	Praseodymium	Pr	59	140.92
Calcium	Ca	20	40.08	Promethium	Pm	61	145
Californium	Cf	98	249.00	Protactinium	Pa	91	231
Carbon	C	6	12.011	Radium	Ra	88	226.05
Cerium	Ce	58	140.12	Radon	Rn	86	222
Cesium	Cs	55	132.91	Rhenium	Re	75	186.22

International Atomic Weights (cont.)

	Symbol	Atomic Number	Atomic Weight		Symbol	Atomic Number	Atomic Weight
Chlorine	Cl	17	35.457	Rhodium	Rh	45	102.91
Chromium	Cr	24	52.01	Rubidium	Rb	37	85.48
Cobalt	Co	27	58.94	Ruthenium	Ru	44	101.10
Copper	Cu	29	63.54	Samarium	Sm	62	150.35
Curium	Cm	96	245	Scandium	Sc	21	44.96
Dysprosium	Dy	66	162.51	Selenium	Se	34	78.96
Erbium	Er	68	167.27	Silicon	Si	14	28.09
Europium	Eu	63	152	Silver	Ag	47	107.880
Fluorine	F	9	19	Sodium	Na	11	22.991
Francium	Fr	87	223	Strontium	Sr	38	87.63
Gadolinium	Gd	64	157.26	Sulfur	S	16	32.066
Gallium	Ga	31	69.72	Tantalum	Ta	73	180.95
Germanium	Ge	32	72.60	Technetium	Tc	43	99
Gold	Au	79	197.00	Tellurium	Te	52	127.61
Hafnium	Hf	72	178.50	Terbium	Tb	65	158.93
Helium	He	2	4.003	Thallium	Tl	81	204.39
Holmium	Ho	67	164.94	Thorium	Th	90	232.05
Hydrogen	H	1	1.0080	Thulium	Tm	69	168.94
Indium	In	49	114.82	Tin	Sn	50	118.70

International Atomic Weights (cont.)

	Symbol	Atomic Number	Atomic Weight		Symbol	Atomic Number	Atomic Weight
Iodine	I	53	126.91	Titanium	Ti	22	47.90
Iridium	Ir	77	192.20	Tungsten	W	74	183.86
Iron	Fe	26	55.85	Uranium	U	92	238.07
Krypton	Kr	36	83.80	Vanadium	V	23	50.95
Lanthanum	La	57	138.92	Xenon	Xe	54	131.30
Lead	Pb	82	207.21	Ytterbium	Yb	70	173.04
Lithium	Li	3	6.940	Yttrium	Y	39	88.92
Lutetium	Lu	71	174.99	Zinc	Zn	30	65.38
Magnesium	Mg	12	24.32	Zirconium	Zr	40	91.22
Manganese	Mn	25	54.94				
Mendelevium	Mv	101	256.00				
Mercury	Hg	80	200.61				
Molybdenum	Mo	42	95.95				

Temperature Conversion Tables

F	*C*	*F*	*C*	*F*	*C*	*F*	*C*	*F*	*C*
-40	-40.0	9	-12.8	58	14.4	107	41.7	156	68.9
-39	-39.4	10	-12.2	59	15.0	108	42.2	157	69.4
-38	-38.9	11	-11.7	60	15.6	109	42.8	158	70.0
-37	-38.3	12	-11.1	61	16.1	110	43.3	159	70.6
-36	-37.8	13	-10.6	62	16.7	111	43.9	160	71.1
-35	-37.2	14	-10.0	63	17.2	112	44.4	161	71.7
-34	-36.7	15	-9.4	64	17.8	113	45.0	162	72.2
-33	-36.1	16	-8.9	65	18.3	114	45.6	163	72.8
-32	-35.6	17	-8.3	66	18.9	115	46.1	164	73.3
-31	-35.0	18	-7.8	67	19.4	116	46.7	165	73.9
-30	-34.4	19	-7.2	68	20.0	117	47.2	166	74.4
-29	-33.9	20	-6.7	69	20.6	118	47.8	167	75.0
-28	-33.3	21	-6.1	70	21.1	119	48.3	168	75.6
-27	-32.8	22	-5.6	71	21.7	120	48.9	169	76.1
-26	-32.2	23	-5.0	72	22.2	121	49.4	170	76.7
-25	-31.7	24	-4.4	73	22.8	122	50.0	171	77.2
-24	-31.1	25	-3.9	74	23.3	123	50.6	172	77.8
-23	-30.6	26	-3.3	75	23.9	124	51.1	173	78.3
-22	-30.0	27	-2.8	76	24.4	125	51.7	174	78.9
-21	-29.4	28	-2.2	77	25.0	126	52.2	175	79.4
-20	-28.9	29	-1.7	78	25.6	127	52.8	176	80.0
-19	-28.3	30	-1.1	79	26.1	128	53.3	177	80.6
-18	-27.8	31	-0.6	80	26.7	129	53.9	178	81.1
-17	-27.2	32	0.0	81	27.2	130	54.4	179	81.7
-16	-26.7	33	0.6	82	27.8	131	55.0	180	82.2
-15	-26.1	34	1.1	83	28.3	132	55.6	181	82.8
-14	-25.6	35	1.7	84	28.9	133	56.1	182	83.3
-13	-25.0	36	2.2	85	29.4	134	56.7	183	83.9
-12	-24.4	37	2.8	86	30.0	135	57.2	184	84.4
-11	-23.9	38	3.3	87	30.6	136	57.8	185	85.0
-10	-23.3	39	3.9	88	31.1	137	58.3	186	85.6
-9	-22.8	40	4.4	89	31.7	138	58.9	187	86.1
-8	-22.2	41	5.0	90	32.2	139	59.4	188	86.7
-7	-21.7	42	5.6	91	32.8	140	60.0	189	87.2
-6	-21.1	43	6.1	92	33.3	141	60.6	190	87.8

Temperature Conversion Tables (cont.)

F	*C*	*F*	*C*	*F*	*C*	*F*	*C*	*F*	*C*
-5	-20.6	44	6.7	93	33.9	142	61.1	191	88.3
-4	-20.0	45	7.2	94	34.4	143	61.7	192	88.9
-3	-19.4	46	7.8	95	35.0	144	62.2	193	89.4
-2	-18.9	47	8.3	96	35.6	145	62.8	194	90.0
-1	-18.3	48	8.9	97	36.1	146	63.3	195	90.6
0	-17.8	49	9.4	98	36.7	147	63.9	196	91.1
1	-17.2	50	10.0	99	37.2	148	64.4	197	91.7
2	-16.7	51	10.6	100	37.8	149	65.0	198	92.2
3	-16.1	52	11.1	101	38.3	150	65.6	199	92.8
4	-15.6	53	11.7	102	38.9	151	66.1	200	93.3
5	-15.0	54	12.2	103	39.4	152	66.7	201	93.9
6	-14.4	55	12.8	104	40.0	153	67.2	202	94.4
7	-13.9	56	13.3	105	40.6	154	67.8	203	95.0
8	-13.3	57	13.9	106	41.1	155	68.3	204	95.6
205	96.1	254	123.3	303	150.6	352	177.8	401	205.0
206	96.7	255	123.9	304	151.1	353	178.3	402	205.6
207	97.2	256	124.4	305	151.7	354	178.9	403	206.1
208	97.8	257	125.0	306	152.2	355	179.4	404	206.7
209	98.3	258	125.6	307	152.8	356	180.0	405	207.2
210	98.9	259	126.1	308	153.3	357	180.6	406	207.8
211	99.4	260	126.7	309	153.9	358	181.1	407	208.3
212	100.0	261	127.2	310	154.4	359	181.7	408	208.9
213	100.6	262	127.8	311	155.0	360	182.2	409	209.4
214	101.1	263	128.3	312	155.6	361	182.8	410	210.0
215	101.7	264	128.9	313	156.1	362	183.3	411	210.6
216	102.2	265	129.4	314	156.7	363	183.9	412	211.1
217	102.8	266	130.0	315	157.2	364	184.4	413	211.7
218	103.3	267	130.6	316	157.8	365	185.0	414	212.2
219	103.9	268	131.1	317	158.3	366	185.6	415	212.8
220	104.4	269	131.7	318	158.9	367	186.1	416	213.3
221	105.0	270	132.2	319	159.4	368	186.7	417	213.9
222	105.6	271	132.8	320	160.0	369	187.2	418	214.4
223	106.1	272	133.3	321	160.6	370	187.8	419	215.0
224	106.7	273	133.9	322	161.1	371	188.3	420	215.6
225	107.2	274	134.4	323	161.7	372	188.9	421	216.1

Temperature Conversion Tables (cont.)

F	*C*	*F*	*C*	*F*	*C*	*F*	*C*	*F*	*C*
226	107.8	275	135.0	324	162.2	373	189.4	422	216.7
227	108.3	276	135.6	325	162.8	374	190.0	423	217.2
228	108.9	277	136.1	326	163.3	375	190.6	424	217.8
229	109.4	278	136.7	327	163.9	376	191.1	425	218.3
230	110.0	279	137.2	328	164.4	377	191.7	426	218.9
231	110.6	280	137.8	329	165.0	378	192.2	427	219.4
232	111.1	281	138.3	330	165.6	379	192.8	428	220.0
233	111.7	282	138.9	331	166.1	380	193.3	429	220.6
234	112.2	283	139.4	332	166.7	381	193.9	430	221.1
235	112.8	284	140.0	333	167.2	382	194.4	431	221.7
236	113.3	285	140.6	334	167.8	383	195.0	432	222.2
237	113.9	286	141.1	335	168.3	384	195.6	433	222.8
238	114.4	287	141.7	336	168.9	385	196.1	434	223.3
239	115.0	288	142.2	337	169.4	386	196.7	435	223.9
240	115.6	289	142.8	338	170.0	387	197.2	436	224.4
241	116.1	290	143.3	339	170.6	388	197.8	437	225.0
242	116.7	291	143.9	340	171.1	389	198.3	438	225.6
243	117.2	292	144.4	341	171.7	390	198.9	439	226.1
244	117.8	293	145.0	342	172.2	391	199.4	440	226.7
245	118.3	294	145.6	343	172.8	392	200.0	441	227.2
246	118.9	295	146.1	344	173.3	393	200.6	442	227.8
247	119.4	296	146.7	345	173.9	394	201.1	443	228.3
248	120.0	297	147.2	346	174.4	395	201.7	444	228.9
249	120.6	298	147.8	347	175.0	396	202.2	445	229.4
250	121.1	299	148.3	348	175.6	397	202.8	446	230.0
251	121.7	300	148.9	349	176.1	398	203.3	447	230.6
252	122.2	301	149.4	350	176.7	399	203.9	448	231.1
253	122.8	302	150.0	351	177.2	400	204.4	449	231.7

Incompatible Chemicals

The substances in the left-hand column must be stored and handled so that they cannot come into any contact with the substances in the right-hand column.

Alakaline and alkaline-earth metals, such as sodium, potassium, cesium, lithium, magnesium, calcium, aluminum	Carbon dioxide, carbon tetrachloride, and other chlorinated hydrocarbons. (Also prohibit water, foam, and dry chemical on fires involving these metals.)
Acetic acid	Chromic acid, nitric acid, hydroxyl-containing compounds, ethylene glycol, perchloric acid, peroxides, and permanganates.
Acetone	Concentrated nitric and sulfuric acid mixtures.
Acetylene	Chlorine, bromine, copper, silver, fluorine, and mercury.
Ammonia (anhydr)	Mercury, chlorine, calcium hypochlorite, iodine, bromine, and hydrogen fluoride.
Ammonium nitrate	Acids, metal powders, flammable liquids, chlorates, nitrites, sulfur, finely divided organics or combustibles.
Aniline	Nitric acid, hydrogen peroxide.
Bromine	Ammonia, acetylene, butadiene, butane and other petroleum gases, sodium carbide, turpentine, benzene, and finely divided metals.
Calcium carbide	Water (See also acetylene.)
Calcium oxide	Water.
Carbon, activated	Calcium hypochlorite.
Copper	Acetylene, hydrogen peroxide.
Chlorates	Ammonium salts, acids, metal powders, sulfur, finely divided organics or combustibles.
Chromic acid	Acetic acid, naphthalene, camphor, glycerol, turpentine, alcohol, and other flammable liquids.
Chlorine	Ammonia, acetylene, butadiene, butane and other petroleum gases, hydrogen, sodium carbide, turpentine, benzene, and finely divided metals.

Chlorine dioxide	Ammonia, methane, phosphine, and hydrogen sulfide.
Fluorine	Isolate from everything.
Hydrocyanic acid	Nitric acid, alkalis.
Hydrogen peroxide	Copper, chromium, iron, most metals or their salts, any flammable liquid, combustible aniline, nitromethane.
Hydrofluoric acid, anhydrous (hydrogen fluoride)	Aqueous or anhydrous ammonia
Hydrogen sulfide	Fuming nitric acid, oxidizing gases.
Hydrocarbons (benzene, butane, propane, gasoline, turpentine, etc.)	Fluorine, chlorine, bromine, chromic acid, sodium peroxide.
Iodine	Acetylene, anhydrous or aqueous ammonia.
Mercury	Acetylene, fulminic acid, ammonia.
Nitric Acid (conc)	Acetic acid, aniline, chromic acid, hydrocyanic acid, hydrogen sulfide, flammable liquids, flammable gases, and nitritable substances.
Nitroparaffins	Inorganic bases.
Oxygen	Oils, grease, hydrogen, flammable liquids, solids or gases.
Oxalic acid	Silver, mercury.
Perchloric acid	Acetic anhydride, bismuth and its alloys, alcohol, paper, wood, grease, oils.
Peroxides, organic	Organic or mineral acids; avoid friction.
Phosphorus (white)	Air, oxygen.
Potassium chlorate	Acids (See also chlorate.)
Potassium perchlorates	Acids (See also perchloric acid.)
Potassium permanganate	Glycerol, ethylene glycol, benzaldehyde, sulfuric acid.
Silver	Acetylene, oxalic acid, tartaric acid, fulminic acid, ammonium compounds.
Sodium	See alkaline metals.
Sodium nitrate	Ammonium nitrate and other ammonium salts.

Sodium oxide	Water.
Sodium peroxide	Any oxidizable substance, such as ethanol, methanol, glacial acetic acid, acetic anhydride, benzaldehyde, carbon disulfide, glycerol, ethylene glycol, ethyl acetate, methyl acetate, and furfural.
Sulfuric acid	Chlorates, perchlorates, permanganates.
Zirconium	Prohibit water, carbon tetrachloride, foam, and dry chemical on zirconium fires.

Safety in the Laboratory or Home Workshop

It is necessary to learn:

- Use of laboratory fume hoods
- Handling flammable solvents
- Mixing acids
- Glass blowing
- Common electrical hazards
- First aid (for four situations only)
  - Stoppage of breathing
  - Profuse bleeding
  - Chemical burns (water only)
  - Fire in clothing
- Use of portable fire extinguishers
- Compressed or flammable gases
- Handling and storing dangerous chemicals, including alkali metals.

An outstanding deficiency pertaining to laboratory safety seems to be a lack of awareness of hazards among nontechnical personnel. It is conceivable that increased emphasis on "briefing" custodial workers about the dangers of the laboratories in which they work, and periodic review of these conditions could substantially reduce the hazard of ignorance.

Third, a more universal use of safety glasses, reaction shields, and other personal protective devices seems to be needed. From the responses received, an increased program of education on the hazards of common laboratory procedures and the use of personal protective equipment to lessen these hazards would be helpful.

Chemical Hazards

All laboratories, whether they be biological, chemical, or radiological, utilize hazardous chemicals. The hazard may result from utilizing the "raw" product or from products of a chemical reaction between two or more substances or breakdown products developed through heating or aging. Laboratory personnel should have an acquaintance at least with the modes of entry, the physiological responses, both acute and chronic, and methods of roughly assessing the hazards potential of chemicals they are using.

Electrical Hazards and Management

The problem of handling electricity is probably one of the most ignored facets of safety, yet each year many needless deaths and injuries are caused through carelessness in handling even low voltages. It is also of importance to recognize that electrical equipment can act as an ignition source to activate a fire or explosion. Static electricity should be considered in this category.

Pressure Hazards

Pressure equipment, either high or low (vacuum), is a part of most laboratories. High-pressure apparatus such as gas cylinders, if improperly handled, can be very dangerous. This is especially true of oxygen. Precautions are necessary in handling, transporting, and in storing. Vacuum equipment, through implosions, can be every bit as dangerous as high-pressure explosions.

Cryogenic Hazards

Cryogenics or the use of low-temperature refrigerants requires a knowledge of the behavior of these materials under laboratory conditions. It is impossible to understand the design of a piece of cryogenic equipment or cryogenic experiment without an appreciation for the principles of insulation or the significance of extremely low temperatures. Misuse can result in severe injury.

Flammable Chemicals-Hazards

Fires and explosions account for the most dangerous and the most expensive types of laboratory accidents. A knowledge of the flammable properties of chemicals along with an understanding of potential sources of ignition is extremely vital. Storage and handling of these materials also requires special attention.

General Safety Considerations

A number of accidents and injuries in laboratories could very well result from improper lifting, falls, and lacerations from improper handling of glassware. Preventive measures in these areas are worthy of mention.

Ventilation

The principal method of hazards control in laboratory involves the effective use of ventilation, both general and exhaust. An example of exhaust ventilation is the fume hood which if improperly designed or used fails to give the desired protection. Observations indicate that the function of this equipment is not entirely understood and a number of misuses have been witnessed.

Laboratory Sanitation

Poor laboratory sanitation practices may be the cause of contaminating potable water supplies through temporary cross connections. At times, poor housekeeping practices may be dangerous because of blocking passages or by providing tripping hazards. Many chemicals are kept well beyond their usefulness, causing containers to deteriorate and leak, or chemicals to become unstable. Disposal of flammable and toxic chemicals also presents a problem.

Protective Equipment

All laboratories require that protective equipment of one type or another be immediately available. These devices may include eye wash, emergency shower, safety glasses, eye shields, protective clothing, and respiratory protection. Knowledge of the proper usage and limitations of such equipment is extremely important. At times, injury or death may result from improper selection and application of protective equipment.

Reports and Records

Reports and records are necessary adjuncts to any safety program and should be complete, accurate, and disseminated to the appropriate administrators. Accident reports are of little value unless periodically examined and tabulated in order to obtain a picture of local and overall problems.

Emergencies

The initial procedures one follows in an emergency oftentimes determine the ultimate outcome of the accident, both to the individuals and to the installation. The rudiments of first aid, fire fighting and reporting are vital. Personnel have to be continually instructed on procedures for medical and file emergencies and how and where to make these initial contacts. Such procedures are critical, especially when working alone.

Contact Lenses

It is important to wear eye protection in the chemical laboratory and expecially when wearing contact lenses.

Danger in Handling Acid

The heat evolution of the solution could have provided sufficient thermal shock to the glass to permit it to crack when lifted free of the counter, or setting it on a cold counter top; or a shock in setting it down, could have contributed to the bottom separating when the jug was lifted.

Written procedures for handling of acids should always be followed. Personal protective equipment consisting of face protection, rubber apron, and gloves are a necessity for this operation.

You Must Have Fire Extinguishers

If a fire breaks out in your office or apartment, get out fast. Many people are killed because they don't realize how fast a small fire can spread.

If you are caught in smoke take short breaths, breathe through your nose, and crawl to escape. The air is better near the floor.

Head for stairs – not elevator. A bad fire can cut off the power to elevators. Close all doors and windows behind you.

If you are trapped in a smoke-filled room, stay near the floor, where the air is better. If possible, sit by a window where you can call for help.

Feel every door with your hand. If it's hot don't open. If it's cool, make this test: open slowly and stay behind the door. If you feel heat or pressure coming through the open door, slam it shut.

If you can't get out, stay behind a closed door. Any door serves as a shield. Pick a room with a window. Open the window at the top and bottom. Heat and smoke will go out the top. You can breathe out the bottom.

DON'T fight a fire yourself.

DON'T jump. Many people have jumped and died without realizing rescue was just a few minutes away.

If there is a panic for the main exit, get away from the mob. Try to find another way out. Once you are safely out, ***DON'T*** go back in. Call the Fire Department immediately. Use alarm box or telephone. DIAL 911

If you find smoke in an open stairway or open hall, use another pre-planned way out.

REMEMBER: Get out fast. Don't underestimate how fast a small fire can spread. Use stairs, not the elevator. Close all doors behind you. Don't panic. Once you are safely out, call the Fire Department. Dial 911 or use alarm box. Don't go back in.

Trademark Chemical Manufacturers

The following is a list of Trademark Chemical Manufacturers. The names of these manufacturers are preceded by a number. In the list of Trademark Chemicals that follows this one, the manufacturers are referred to by a number alone.

1. Advance Solvents .New Brunswick, N.J.
2. Air Products & Chemicals, Inc.. . . . . .Philadelphia, Pa.
3. Air Reduction Chem & Carbide Co. . . . . .New York, N.Y.
4. Albright & Wilson, Ltd. . . . . England
5. Allied Chem. Corp. . . . . Morristown, N.J.
6. Allied Colloids, Ltd.. . . . . England
7. Aluminum Co. of Amer. . . . . Pittsburgh, Pa.
8. American Cyanamid Co. . . . . .Wayne, N.J.
9. American Lecithin Corp.. . . . . .Woodside, L.I., N.Y.
10. American Mineral Spirits. . . . . .New York, N.Y.
11. American Zinc Sales. . . . . .Columbus, Ohio
12. Archer Daniels Midland Co.. . . . . .Decatur, Ill.

12A. Argus Chem. Corp. . . . . Brooklyn, N.Y.

13. Ashland Chem. Co. . . . . .Columbus, Ohio
14. Baker Castor Oil . . . . Bayonne, N.J.
15. BASF Corp. . . . . Paramus, N.J.
16. Berol Kemi (UK) Ltd. . . . . England
17. Borden Chem. Co. . . . . .New York, N.Y.

18. British Celanese Ltd. . . . . England
19. BTP Tioxide. . . . . England
20. Buckman Labs . . . . .Memphis, Tenn.
21. Burgess Pigment Co.. . . . . .Sandersville, Ga.
22. Byk Mallinckrodt . . . . .Rahway, N.J.
23. Cabot Corp. . . . . Boston, Mass.
24. Carbide & Carbon Chemicals Co. . . . . .New York, N.Y.
25. Cargill, Inc. . . . . .Minneapolis, Minn.
26. Celanese Resins. . . . . .New York, N.Y.
27. Ciba-Geigy Corp.. . . . . Ardsley, N.Y.
28. Cities Service . . . . Atlanta, Ga.
29. Colloids, Inc. . . . . .Newark, N.J.
30. Columbian Carbon Co., Inc. . . . . .New York, N.Y.
31. Columbian Div. of Cities Service Co. . . . . Atlanta, Ga.
32. Commercial Solvents Corp. . . . . .New York, N.Y.
33. Cosan Chem. Co.. . . . . Clifton, N.J.
34. Joseph Crosfield & Sons, Ltd. . . . . England
35. Croxton & Garry Ltd. . . . . England
36. Crucible Chem. Co. . . . . Greenville, S.C.
37. Davidson Chem. Co.. . . . . .Baltimore, Md.
37A. Day-Glo Color Corp. . . . . Cleveland, Ohio
38. Degussa, Inc. Chem. Div.. . . . . Teterboro, N.J.
39. De Lore Div. of NL Ind. . . . . .New York, N.Y.
40. Dewey & Almy Chem. Div. of W.R. Grace & Co. . . Cambridge, Mass.
41. Dexter Chem. Corp.. . . . . Bronx, N.Y.
42. Diamond Shamrock Chem. Co. . . . . Cleveland, Ohio
43. Dow Chem. Co.. . . . . .Midland, Mich.
44. Dow Corning Corp. . . . . .Midland, Mich.
45. Drew Chem. Corp.. . . . . Boontown, N.J.
46. E.I. du Pont de Nemours & Co. . . . . Wilmington, Del.
47. Eagle Picher Ind., Co. The. . . . . .Cincinnati, Ohio
48. Eastman Chem. Prod., Inc. . . . . Kingsport, Tenn.
49. English Mica. . . . . England
50. Exxon Co. . . . . Houston, Tex.
51. Ferro Corp. . . . . Cleveland, Ohio
52. Firestone Plastics. . . . . .Pottstown, Pa.
53. Firestone Rubber & Synthetic Latex Co. . . . . Akron, Ohio
54. FMC Corp.. . . . . .New York, N.Y.
55. Fordamin Co. Ltd.. . . . . England

56. Freeport Kaolin Co.. . . . . .New York, N.Y.
57. GAF Corp.. . . . . .New York, N.Y.
58. Garrick Chem. & Equip. Co., Ltd.. . . . . England
59. General Electric . . . . Waterford, N.Y.
60. General Mills Corp. Chem. Div. . . . . Minneapolis, Minn.
61. Georgia-Kaolin Co. . . . . .Elizabeth, N.J.
62. Glidden-Durkee Div. of SCM Corp. . . . . Jacksonville, Fla.
63. Glyco Chem. Co.. . . . . .Greenwich, Conn.
64. B.F. Goodrich Chem. Co. Cleveland, Ohio
65. Goodyear Tire & Rubber Co.. Akron, Ohio
66. W.R. Grace & Co. . . . . .New York, N.Y.
66A. Great Lakes Carbon Corp.. . . . . .New York, N.Y.
67. Harshaw Chem. Co., The. Cleveland, Ohio
68. Hercules Chem.. . . . . Wilmington, Del.
69. Hercules Powder Co. . . . . Wilmington, Del.
70. Hercules Powder Co., Ltd.. . . . . England
71. Hooker Chem. Corp. . . . . Niagara Falls, N.Y.
72. J.M. Huber Corp.. . . . . Edison, N.J.
73. Humble Oil and Refining Co.. . . . . .New York, N.Y.
74. ICI America Inc. . . . . Wilmington, Del.
75. Imperial Chem. Ind., Ltd., Org. Div. . . . . England
76. Imperial Color Chem.. . . . . Glenn Falls, N.Y.
77. Industrial Minerals. . . . . .Canada
78. International Talc Co. . . . . .New York, N.Y.
79. Jennat Corp.. . . . . Torrance, Calif.
80. Johns-Manville . . . . .New York, N.Y.
81. Jones-Dabney Div. of Devoe & Reynolds Co., Inc.. . . . Louisville, Ky.
82. Kelco Corp.. . . . . Clark, N.J.
83. M.W. Kellogg Co.. . . . . .New York, N.Y.
84. Koppers Co.. . . . . .Pittsburgh, Pa.
85. Kronos Titanium Pigments Ltd. England
86. Lanson Chem. Co.. . . . . .East St. Louis, Ill.
87. Laporte Ind., Ltd.. . . . . England
88. M & T Chemicals. . . . . .Rahway, N.J.
89. Marbon Chem. Div. of Borg Warner Corp. . . . .Washington, W. Va.
90. McClosky Varnish Co. . . . . .Philadelphia, Pa.
91. McWhorter Chem. Co. Div. . . . . .Chicago, Ill.
92. Melbourne Chem. Ltd.. . . . . England
93. Merck & Co.. . . . . .Rahway, N.J.

94. Metalsalts Corp. . . . . .Hawthorne, N.J.
95. Microfine Minerals & Chem. Ltd. . . . . England
96. Minerals & Chem. Div. of Phillipp Corp.. . . . . Edison, N.J.
97. Mobay Chem. Co. . . . . .Pittsburgh, Pa.
98. Mobil Chem. Co.. . . . . .New York, N.Y.
99. Monsanto Chem. Co. . . . . St. Louis, Mo.
100. National Lead Co. . . . . Hightstown, N.J.
101. National Starch & Chem. Co.. . . . . .Plainfield, N.J.
102. New Jersey Zinc Co. . . . . .Palmerton, Pa.
103. Nopco Chem. Co. . . . . .Newark, N.J.
104. Norwegian Talc Ltd. . . . . England
105. Nuodex Div. of Tenneco Chem. Inc. . . . . .Piscataway, N.J.
106. Peerless Chem. Co.. . . . . Detroit, Mich.
107. Penn. Ind. Chem. Co.. . . . . Clairton, Pa.
108. Penn. Glass & Sand Corp. . . . . .Pittsburgh, Pa.
109. Charles Pfizer . . . . .New York, N.Y.
110. Polyvinyl Chem. . . . . Wilmington, Mass.
111. Refined Onyx Div. of Millmaster Onyx Corp. . . . . Lyndhurst, N.J.
112. Raybo Chem. Co. . . . . .Huntington, W. Va.
113. Reheis Chem. Co. . . . . Phoenix, Ariz.
114. Reichhold Chem. Inc.. . . . . White Plains, N.Y.
114A. Resyn Corp.. . . . . Linden, N.J.
115. Rohm & Haas . . . . .Philadelphia, Pa.
116. Rohm & Haas Ltd.. . . . . England
117. Ross & Rowe, Inc.. . . . . .New York, N.Y.
118. Scott Bader & Co., Ltd. . . . . England
119. Shell Chemicals (UK) Ltd.. . . . . England
120. Shell Oil Co.. . . . . .New York, N.Y.
121. Sherwin Williams Chem. . . . . Cleveland, Ohio
122. Spencer Kellogg Div. of Textron, Inc. . . . . .Buffalo, N.Y.
123. Staley Chem. Co.. . . . . .Decatur, Ill.
124. Standard Oil Co. of Calif. . . . . .San Francisco, Calif.
125. Stauffer Chem. Co. . . . . .Westport, Conn.
126. Tamms Ind. . . . . .Chicago, Ill.
127. Thibaut & Walker . . . . .Newark, N.J.
128. Thompson, Weinman & Co.. . . . . Cartlesville, Ga.
129. Thor Chem. Ltd.. . . . . England
130. Titanium Pigment Corp.. . . . . So. Amboy, N.J.
131. Troy Chem. Corp.. . . . . .Newark, N.J.

133. Ultra Adhesives Inc. . . . . Paterson, N.J.
134. Union Carbide Corp. . . . . New York, N.Y.
135. Union Carbide Chem. & Plastics Div. . . . . New York, N.Y.
136. Unitech Chem. Mfg. Co. . . . . Sun Valley, Calif.
137. R.T. Vanderbilt. . . . . Norwalk, Conn.
138. Velsicol Co. . . . . Chicago, Ill.
139. Victor Chem. Co. . . . . Chicago, Ill.
140. Vinyl Prod. Ltd. . . . . England
141. Whittaker, Clark & Daniels . . . . So. Plainfield, N.J.
142. Witco Chem. Co. . . . . New York, N.Y.
143. Wyandotte Chem. Corp. . . . . Wyandotte, Mich.

Trademark Chemicals

A-4436 76
Abitol 68
Acicular # 417 47
Acronal 15
Acryloid 115
Acrysol 115
Acticide 129
Additive # 11 44
Advacar 1
Aerosol 38
Aircoflex 2
Albacar 63
Alcoa 7
Alloprene 74
Al-Sil-Ate 56
Altowhite 61
Amberlac 115
Amsco Solvent B 10
Antiterra U 22
Aquablak 30
Arochem 12
Aroclor 99
Aroflint 13
Arolon 12
Aromatic 100 50
Aroplaz 13
Asbestine 78
ASP-100, -103, -170 96
Astra-Glaze 61
Atomite 128
Aurasperse 67
AZO 11

Bakelite 135
Balab 142
Barytes 39
Beckamine 114
Beckosol 114
Beetle 8
Ben-A-Gel 100
Benax 43
Bentone 100
Bentonite 100
Bermocoll 16
BP-3660 46
Busan 20
Butle 8
Butrol 20

Butyl Dioxitol 119
BYK-P-104-3 22

Cab-O-Sil 23
Calgon. 4
Carbitol. 135
Carbowax 134
Cargill 25
Castung 14
Catalyst 1010 8
Catalyst SC-3900. 59
Celacol 18
Celacose 18
Celite 80
Cellosize 134
Cellosolve 134
Cerechlor. 74
Chemacoil 91
Chevron #3, #225. 124
Chlorowax 42
Colloid 581-B, 600, 606,
643, 677 29
Cosan 33
Courlose 18
Cycopol. 84
Cymel 8
Cyplex 8

D Powder. 55
Dalpad 43
Daxad 40
Day-Glo 37A
Dee-Fo 133
Desmodur 97
Dicalite 66A
DMP-30. 115
Dow 43
Dowanol 43
Dowicil 43
Drapex 12A
Drew L-475 45
Drikalite 128
Duramite 128
Duraplex 115

EKS-2002 134
Ektasolve. 48
Empilan. 4
Emultex 123
Epi-Cure 26
Epi-Rez 26
Epon. 119
ERL-2774 134
Estynox. 14
Everflex. 66
Exkin 105
Exon. 52

FCR-1261 53
Ferro 51
Ferrophos 71
FF-30 17
Fibrene 136
Flexbond. 3
Flexol 24
Foamaster 42
Foamkill 36
Fordacol 55
Freon 46
FRN-505 53

G-Cure 60
Genepoxy 60
Geon. 64
Glomax 106
Gold Bond R 126
Granicalcium 35

Hercules 357 69
Hi-Sol 13
Hydragloss 72
Hydrite 61
Hypalon 46

IB-1550 25
Icecap K 21
Igepal 57
Imperial X-2806 68
International 3X 78
Ionol 120
Isopar 50

Kadox 102
Kaopaque 61
Kelecin 122
Kelzan 83
Kodaflex 48
Kronos 85

L-5310 134
Lankyd 86
Latekoll 15
Lexinol 9
Lo Micron 399 141

Mapico 30
Marbon 89
Mark 12A
Melamine 115
Merbac 93
Metasol 94
Methocel 43
Mica 325 Mesh 49
Microcal 34
Microdol 104
Microfine 95
Micromica 104
Minex 77
Min-U-Sil 108
Mobilar 98
Modaflow 99
Molywhite 121
Mondur 97
Monoplex 116
M-P-A 60 14
Multiflex 42
Multron 97

N Lecithin 25
Nacconate 5
Natrosol 70
Neocryl 110
Neosol 119
Neospectra 31
Nopco 103
Nopcosant 103
Nuodex 105
Nuosperse 105
Nuxtra 105
Nytal 137

Omya 92
Oncor 100
OR-580 8

ParCryl 127
Paralac 73
Paraplex 116
Parco 71
Parez 8
Parlon 69
Peerless 28
Pentalyn G 68
Perma-Kleer 111
Phos-Chek 99
Piccodiene 107

PKHH 134
Pliolite 65
PMA-30. 33
PMD-11 105
PMO-30. 134
Polidene 118
Polyamide 815X-70 60
Polyco. 17
Polyketone. 134
Polymul. 103
Poly-Tex 81
Primex 39
Propasol B 134
Proxel 75
Purecal 143

R-900 46
R-2199 109
Raven 1035 28
Raybo. 112
Rayox. 137
RCI 93-905 114
Regal 23
Resimene. 99
Resyn 1255 79
Resyn 2243 101
Resypox 114A
Rezyl 84
Rhoplex 115
Runa. 87

SAG-470 115
Santicizer. 99
Santocel 99
Satin-White 68
Silbond 125
Skane 115
Snowflake 128
Solvesso. 73

Sparmite 109
SR-112 59
Staybelite 68
Stayco. 123
Strodex 41
Super Ad-It 105
Surfactol 13 14
Surfex. 42
Syloid 37
Syntex 26
Syn-U-Tex 81

Tamol 115
Tanblok. 113
Tecsol 48
Texanol. 48
Thermolite. 88
Thickener L 57
Thixatrol 14
Thixcin 14
Ti Pure 46
Tinuvin 27
Tioxide 19
Titanox 130
Tri Mal 100
Triton 115
Troykyd 32
Troysan. 131
TSPP. 54

Ubatol. 123
Ucar 134
Ultra White 96
Unitane 8
United 3017L. 13

VAGH Resin 134
Van-Amid 137
Vansolve 137

Varcopol 90
Veegum 137
Velsicol 138
Veroc 141
Versamid 60
Vicron 70
Victawet 139
Vinacryl 17
Vinapol 140
Vinoflex 15
Viscalex 6

Wallkyd 114
Wallpol 114
325 WG 49
Winnofil 74
Witcamide 142

X-2280 69
X5R 39
XAMA-7 58

Y539D 46
Yelkin 117

Zanflo 82
Zeolex 72
Zopaque 62

INDEX

Acid, danger in handling, 378
Acrylic, aircraft, nitrocellulose, 140-141
Acrylic automotive topcoat, clear, 306
Acrylic baking enamel, general purpose, 341-342
  epoxy white, 344
Acrylic can enamel, exterior, 142-143
Acrylic can coating, interior, 143-144
Acrylic cloth coating, 315
Acrylic corrosion-inhibiting primer, 8-11
Acrylic decorative coatings, 307
Acrylic dry-ground deep tone paint, 269-272
Acrylic emulsion paint, exterior, 58-61
Acrylic enamel, aluminum, 179
  color-retentive, general purpose, 180
  gloss black, 183
  gloss white, 181-182
  interior, 260-263
  latex, 190-192
  tintable eggshell, 183-184
  water reducible, 184-190
  white spray, 212-213
Acrylic exterior paint, deep tone, 104-109
  latex, 62-63
  latex white, 84-85
  white, 74-84
Acrylic exterior primer, 32-33
Acrylic exterior topcoat, white, 120-123
Acrylic exterior trim paint, 126-136
Acrylic flat wall paint, 217-222
Acrylic flexible vinyl substrate topcoat, 306
Acrylic house paint, 61-62
Acrylic industrial enamel, black, 339
  gloss, 340
  white, 337-339
Acrylic industrial hardboard primer, 31
Acrylic ink, gravure, 316
  polyvinylidene film, 317
Acrylic interior topcoat, white, 241-252
Acrylic lacquer, aqueous, 290-292
  automotive, gray, 141
  black, 293
Acrylic-latex metal primer, 37
Acrylic-latex swimming pool paint, 322-323

Acrylic-latex traffic paint, 325-328
Acrylic masonry floor paint, 167-171
Acrylic metal paint, aerosol, 149-150
Acrylic-modified alkyd spray enamel, 214
Acrylic tint base, economy, 273-274
  exterior sheen, 114-115
  exterior, light, for glycol dispersed colorants, 112-114
  interior, eggshell, 276
  interior, flat, universal, 275-276
  interior semigloss for water and glycol dispersed colorants, 272-273
Acrylic tradesales brushing varnish, 298
Acrylic urethane clear topcoat, 309
Acrylic urethane coating, fluorescent, 309
Acrylic urethane enamel, white, 192
Acrylic wood finish, 299
Aerosol enamel, alkyd, 214-215
Aerosol paint, metal, acrylic, 149-150
Aircraft paint, acrylic-nitrocellulose, 140-141
Air-dry wood lacquer, thermoplastic acrylic/butyrate, 294
Airless spray, alkyd metal paint, 150
Alkyd aluminum paint, 151-152
Alkyd bake enamel, 345-349
  blue metallic, 352
  gloss, 350-351
  high gloss, yellow, 354
  medium oil, 351-352
  short oil, 350
Alkyd-chlorinated rubber traffic line paint, 329
Alkyd corrosion-inhibiting primer, 11-12
Alkyd electrocoating black enamel, 340-341
Alkyd enamel, 172-177
  acrylic-modified, 214
  aerosol, 214-215
  black, 199-200
  blue, 198-199
  chrome yellow, 202
  coil coating, 145
  electroplating, 340-341
  floor, 180
  interior, 263-264
  interior orange, 201
  interior red, 197-198
  interior toluidine red, 201
  interior white, 192-197
Alkyd flat wall paint, 222-224
Alkyd interior finish coat, white, 252
Alkyd metal paint, 152-153
  airless spray, 150
  aluminum, 151-152
Alkyd modified flexible polyester wood coating, 299-300
Alkyd overprint varnish, 297
Alkyd primer, 37-41
  corrosion-inhibiting, 11-12
  spray, 6
  zinc chromate, 16-17
Alkyd screen process color, 322
Alkyd shingle stain, red, 137-138
Alkyd traffic line paint, 330
  chlorinated rubber, 329
Alkyd weldable metallically pigmented coating, 153
Aluminum enamel, acrylic, 179
Aluminum paint, alkyd, 151-152
Anticorrosive styrene-acrylic primer, 6-7
Automotive clear topcoat, acrylic, 306
Automotive lacquer, acrylic, gray, 141

Automotive vinyl top-texturizer, 306-307

Bake enamel, acrylic-epoxy, white, 344
alkyd, 345-349
alkyd, blue metallic, 352
alkyd high gloss, yellow, 354
alkyd medium oil, 351-352
alkyd oil-free, 349-350
alkyd short oil, 350
alkyd, white, 345-349
epoxy, and primer, 353
epoxy, white, 354
high gloss, white, 344-345
industrial, polyester, high gloss, white, 355
water reducible, high gloss, white, 343-344
Baking primer, epoxy, gray, 16
Board coating, latex, 313-314
vinyl acrylic, satin-finish, 313
Butyrate-acrylic wood lacquers, 289
Butyrate-amino paper lacquers, 296

Cal Ink latex tint base, 319-321
Can coating, acrylic, interior, 143-144
epoxy, exterior-interior, 144
phenolic, interior-exterior, 145
Can enamel, acrylic, exterior, 142-143
Casein/latex offset coating, 315
Catalyzed wood topcoat, 300
Ceiling paint, styrene-acrylic, 239-240
interior, 240
wall, 239
Cellulose nitrate paper lacquer, 296-297
Chemical hazards, 376
Chemicals, incompatible, 372-375
Chlorinated rubber primer, 12-16
metal, 41-43
Chlorinated rubber steel paint, 153-154
Chlorinated rubber swimming pool and masonry, 158-159
Chlorinated rubber traffic paint, 330-331
Cloth coating, acrylic, 315
Coating, board, latex, 313-314
vinyl acrylic, satin-finish, 313
Coating, can, 143-145
epoxy, interior-exterior, 144
Coating, epoxy-coal tar, 310
Coating, epoxy, film corrosion resistant, 363-364
maintenance, 357-361
Coating, epoxy-versamid, flat, 361-362
high build, 310-314
Coating, fluorescent, acrylic urethane, 309
Coating, mastic, high build self-priming, 311-312
Coating, offset, casein/latex, 315
Coating, paper, latex, 314
Coating, vinyl, 308-309
Coil coating, enamel, alkyd, 145
polyester, white, 145-148
primer, vinyl, 48
Common indicators, pH ranges, 366
Concrete floor paint, interior, latex, 171
Contact lenses, 378
Corrosion-inhibiting primer, acrylic, 8-11
alkyd, 11-12
Cryogenic hazards, 376

Decorative coatings, acrylic, 307
Deep tone paint, dry-ground, 269-272
exterior, acrylic, 104-109

Dust primer, PVC zinc, 23-24
silicate zinc, 25-27
zinc, epoxy polyamide, 19-20

Electrical hazards, 376
Electrocoating enamel, alkyd, black, 340-341
Electrostatic disc spray enamel, gloss, white, 335
Emergencies, 378
Emulsion paint, ethylene vinyl acetate, 253
exterior acrylic, 58-61
exterior vinyl acrylic, 73-74
interior-exterior, latex, 287-288
Enamel, acrylic, baking, general purpose, 341-342
color-retentive, general purpose, 180
gloss black, 183
gloss white, 181-182
industrial black, 339
industrial gloss, 340
industrial white, 337-339
interior, 260-263
latex, 190-192
modified alkyd spray, 214
spray, white, 212-213
tintable eggshell, 183-814
urethane, white, 192
water reducible, 184-190
Enamel, alkyd, 172-177
aerosol, 214-215
black, 199-200
blue, 198-199
chrome yellow, 202
electrocoating, black, 340-341
floor, 180
interior, 263-264
interior orange, 201
interior toluidine red, 201
red, 197-198
white, 192-197
Enamel, aluminum, acrylic, 179
Enamel, baking, acrylic-epoxy, white, 344
alkyd, blue metallic, 352
alkyd gloss, 350-351
alkyd high gloss yellow, 354
alkyd, medium oil, 351-352
alkyd, short oil, 350
alkyd, white, 345-349
epoxy, 353-354
epoxy, white, 354
high gloss, white, 344-345, 355
industrial, polyester high gloss, white, 355
oil-free, 349-350
water reducible, high gloss, 343-344

Enamel, can, exterior, acrylic, 142-143
Enamel, coil coating, alkyd, 145
Enamel, epoxy, 202-203
epoxy-polyamide, 203-205
epoxy-polyester, 205-206
epoxy, solventless, thick film, white, 256-257
epoxy-urea/formaldehyde, 206-207

Enamel for electrostatic disc spray, gloss white, 335
Enamel, interior, industrial, 335-356
Enamel, latex, eggshell, 210
latex interior, 265-269
latex interior, flat, 177-178
latex, white, 208-210
Enamel, maintenance, triamino ester, chrome yellow, 364
Enamel, metallized, high gloss polyester, 178
Enamel, polyester, spray, 215-216
white, 211, 215-216
Enamel, polyurethane, gloss white, 179

Enamel tint base, white or light, latex semigloss, 177
Enamel, triamino ester, white, 211-212
Epoxy baking enamel and primer, 353
Epoxy baking enamel, white, 354
Epoxy baking primer, gray, 16
Epoxy can coating, interior-exterior, 144
Epoxy coal tar coating, 310
Epoxy corrosion resistant coating (film), thick, 363-364
Epoxy enamel, 202-203
  solventless, thick film, white, 256-257
Epoxy ester floor varnish, 297
Epoxy maintenance coating, 357-361
Epoxy maintenance paint, exterior, white, 85-86
Epoxy-polyamide enamel, 203-205
Epoxy-polyamide zinc dust primer, 19-20
Epoxy-polyester enamel, 205-206
Epoxy primer, zinc-rich, 17-19
Epoxy thermoplastic ink, 317
Epoxy-urea/formaldehyde enamel, 206-207
Epoxy-versamid coating, high build, 310-311
  flat, 361-362
Epoxy-versamid gel coat, 362-363
Equipment, protective, 377
Ethylene vinyl acetate emulsion paint, 253
Ethyl vinyl acetate exterior paint, 63-64
Exterior acrylic deep tone paint, 104-109
  barn red, 104-105
  black, 108-109
  dark brown, 106-108
  gold yellow, 105
  green, 106
Exterior acrylic emulsion paint, 58-61
Exterior acrylic tint base, light, for glycol dispersed colorants, 112-114
  sheen, 114-115
Exterior acrylic trim paint, 126-136
Exterior acrylic white paint, 74-84
Exterior coating, styrene-acrylic, 71
Exterior-interior flat paint, vinyl acetate, 240
Exterior latex paint, 64-69
  masonry, 157-159
  white, 86-93
Exterior latex tint, white or light, 108-111
Exterior latex tint base, light, 119
  semigloss, 115-119
Exterior latex topcoat, chalking white, 125
  pigmented, 125-126
Exterior latex trim paint, 136-137
Exterior latex wood paint, 159
Exterior paint, acrylic latex, 62-63
  acrylic latex white, 84-85
  ether vinyl acetate, 63-64
  polyvinyl acetate, 69-71
  polyvinyl acetate, with triamino ester, 93-99
  vinyl acrylic, 72-73
Exterior primer, acrylic, 32-33
  latex, 34-36
Exterior tint base, vinyl ethylene, 120
Exterior topcoat, acrylic white, 120-123
Exterior vinyl acetate textured finish, 139
Exterior vinyl acrylic emulsion paints, 73-74

interior, PV acetate, 235
interior, PV acetate, white, 259
interior, styrene-acrylic, 237-239
white paint, 100-102
white topcoat, 124-125
Exterior vinyl chloride-acrylic house paint, white, 102-104
Exterior vinyl ethylene tint, white and light, 111-112

Fabric coating, 307-308
Finish coat, alkyd, interior white, 252
Fire extinguishers, 378-379
Flame retardant paint, polyvinyl chloride, 303-305
Flammable chemicals, hazards, 376-377
Flat paints, exterior-interior, vinyl-acetate, 240
interior latex white, 253-259
Flat tint base, universal, acrylic, interior, 275-276
vinyl acrylic, 283-284
Flat wall paint, acrylic, 217-222
alkyd, 222-224
latex, 224-235
PV acrylic, interior, 235-237
vinyl-acrylic, interior, 241
Floor enamel, alkyd, 180
Floor paint, concrete, interior, latex, 171
masonry, acrylic, 167-171
Floor varnish, epoxy ester, 297
Fluorescent acrylic urethane coating, 309
Fluorescent screen ink, general purpose, 316
Fluorescent silk-screen ink vehicle, 316

Gel coat, epoxy-versamid, 362-363
General purpose baking enamel, acrylic, 341-342
General purpose phosphorescent, 310
Gloss enamel, electrostatic disc spray, white, 335
industrial, acrylic, 340
Gloss white flexible butyrate lacquer, 294-295
Gravure ink, acrylic, 316
styrene-alkyd, 318

Hardboard primer, industrial, acrylic, 31
Hazards, 376-377
chemical, 376
cryogenic, 376
electrical, 376
flammable chemicals, 376-377
pressure, 376
High build coating, epoxy-versamid, 310-311
mastic, self-priming, 311-312
Home workshop safety, 375-379
House paint, acrylic, 61-62
vinyl chloride-acrylic, exterior, white, 102-104

Incompatible chemicals, 372-375
Industrial enamel, acrylic, black, 339
acrylic gloss, 340
acrylic white, 337-339
baking, polyester, high gloss white, 355
interior, 355-356
Industrial hardboard primer, acrylic, 31
Ink, epoxy thermoplastic, 317
gravure, acrylic, 316
gravure, styrene-alkyd, 318
maleic, thermoplastic, 313

polyvinylidene film, acrylic, 317
screen, fluorescent, 316
urea, thermoplastic, 318-319
vehicle, silkscreen fluorescent, 316
Interior ceiling paint, styrene-acrylic, 240
Interior enamel, acrylic, 260-263
alkyd, 263-264
alkyd, orange, 201
alkyd toluidine red, 201
latex, 265-269
latex, flat, 177-178
Interior-exterior emulsion paint, latex, 287-288
Interior finish coat, alkyd, white, 252
Interior flat paint, latex, white, 253-259
polyvinyl acetate, 259
PV acetate, 235
styrene-acrylic, 237-239
Interior flat wall paint, PV acrylic, 235-237
vinyl-acrylic, 241
Interior industrial enamel, 355-356
Interior primer sealer, 1-2
Interior textured finish, styrene-acrylic, 284-286
Interior tint base, acrylic, eggshell, 276
acrylic, semigloss for water and glycol dispersed colorants, 272-273
acrylic, universal, flat, 275-276
Interior topcoat, acrylic, white, 241-252
Interior wall primer, 5
International atomic weights, 367-369

Laboratory safety, 375-379
Laboratory sanitation, 377
Lacquer, acrylic, aqueous, 290-292
acrylic, black, 293
automotive, acrylic, gray, 141
flexible butyrate, gloss white, 294-295
nitrocellulose, 295-296
nitrocellulose, pigmented, 295
paper, butyrate-amino, 298
paper, cellulose nitrate, 296-297
thermoplastic, acrylic/butyrate, air-dry, 294
wood, butyrate acrylic, 289
Latex board coating, 313-314
Latex enamel, acrylic, 190-192
eggshell, 210
white, 208-210
Latex exterior paint, 64-69
acrylic, 62-63
acrylic white, 84-85
masonry, 157-159
wood, 159
Latex exterior primer, 34-36
Latex exterior tint, white and light, 108-111
Latex exterior tint base, light, 119
semigloss, 115-119
Latex exterior topcoat, chalking white, 125
pigmented, 125-126
Latex exterior trim paint, 136-137
Latex flat wall paint, 224-235
Latex "Hypalon" paint for rigid or semirigid substrates, 336
Latex interior concrete floor paint, 171
enamel, 265-269
exterior emulsion paint, 287-288
Latex interior flat enamel, 177-178
flat paint, white, 253-259
Latex masonry paint, multipurpose, 160-166
stucco, 160

Latex paper coating, 314
Latex primer, maintenance, 50-57
maintenance, styrene-acrylic, 57
metal, styrene-acrylic, 47
red iron oxide zinc chromate, 20-21
zinc oxide, 21-22
Latex stucco paint, 160
Latex swimming pool paint, 323-324
Latex tint base, 277-283
Cal Ink colors, 319-321
semigloss, white or light, 177
Latex traffic line paint, 331-334

Maintenance coating, epoxy, 357-361
Maintenance paint, exterior, epoxy, white, 85--86
styrene-acrylic, 364-365
Maintenance primer, latex, 50-57
styrene-acrylic, latex, 57
Maleic thermoplastic ink, 317
Manufacturer's trademark chemicals, 379-383
Masonry paint, chlorinated rubber, 158-159
floor, acrylic, 167-171
latex, 157-166
latex, exterior, 157-166
latex, multipurpose, 160-166
Mastic coating, high build, self-priming, 311-312
weather barrier, 156-157
Metal paint, 149-155
acrylic, aerosol, 149-150
alkyd, 152-153
alkyd, airless spray, 150
Metal primer, acrylic latex, 37
chlorinated rubber, 41-43
PVC, 44-46
ship, phenoxy, 43
styrene acrylic latex, 47
Metallically pigmented weldable coating, alkyd, 153
Metallized enamel, polyester, high gloss, 178

Nitrocellulose lacquer, 295-296
pigmented, 295

Offset coating, casein/latex, 315
Oil-free alkyd baking enamel, 349-350
Overprint varnish, alkyd, 297

Paper coating, latex, 314
Paper lacquer, butyrate-amino, 296
cellulose nitrate, 296-297
pH ranges of common indicators, 366
pH values, 366
Phenolic can coating, interior-exterior, 145
resin primer, 28-29
Phenoxy primer, ship metal, 43
zinc, 22-23
Phosphorescent, general purpose, 310
Pigmented nitrocellulose lacquer, 295
Polyester baking enamel, industrial, high gloss white, 355
Polyester coil coating, 145-148
Polyester enamel, metallized, high gloss, 178
spray, white, 215-216
white, 211
Polyester wood coating, alkyd modified, flexible, 299-300
unsaturated, 301
clear, force-drying, 300-301
Polyurethane enamel, gloss white, 179

Polyvinyl acetate exterior paint, 69-71
white, with triamino ester, 93-99
Polyvinyl acetate interior flat paint, white, 259
Polyvinyl butyral wash primer, 49
Polyvinyl chloride flame retardant paint, 303-305
Polyvinyl chloride swimming pool paint, 324
Polyvinyl chloride traffic line paint, 334-335
Polyvinylidene film ink, acrylic, 317
Pressure hazards, 376
Primer, alkyd, 37-41
alkyd zinc chromate, 16-17
and epoxy baking enamel, 353
anticorrosive, styrene-acrylic, 6-7
chlorinated rubber, 12-16
coil coating, vinyl, 48
corrosion-inhibiting, acrylic, 8-11
corrosion-inhibiting, alkyd, 11-12
epoxy, baking, gray, 16
epoxy, zinc-rich, 17-19
exterior, acrylic, 32-33
exterior, latex, 34-36
industrial hardboard, acrylic, 31
interior sealer, 1-2
interior, wall, 5
latex red iron oxide, zinc chromate, 20-21
latex zinc oxide, 21-22
maintenance, latex, 50-57
maintenance, styrene acrylic latex, 57
metal, acrylic latex, 37
metal, chlorinated rubber, 41-43
metal, PVC, 44-46
metal, styrene-acrylic latex, 47
phenolic resin, 28-29
phenoxy zinc, 22-23
polyvinyl butyral wash, 49
PVC zinc chromate, 24-25
PVC zinc dust, 23-24
ship metal, phenoxy, 43
silicate zinc dust, 25-27
triamino ester, 29-31
vinyl, 48-49
vinyl acrylic, 3-5
wood, styrene-acrylic, 36-37
Protective equipment, 377
PV acetate interior flat paint, 235
PV acrylic interior flat wall paint, 235-237
PVC metal primer, 44-46
PVC zinc chromate primer, 24-25
PVC zinc dust primer, 23-24

Reports and records, 377
Resin primer, phenolic, 28-29
Rust-inhibitive wash, 139

Safety considerations, laboratory and home workshop, 375-379
chemical hazards, 376
contact lenses, 378
cryogenic hazards, 376
danger in handling acid, 378
electrical hazards, 376
emergencies, 378
fire extinguishers, 378-379
flammable chemicals, 376-379
general considerations, 377
laboratory sanitation, 377
pressure hazards, 376
protective equipment, 377
reports and records, 377
Sanitation, laboratory, 377
Screen ink, fluorescent, 316
Screen process color, alkyd, 322
Sealer, interior primer, 1-2
Shingle stain, alkyd, red, 137-138
Ship metal primer, phenoxy, 43
Silicate zinc dust primer, 25-27

Silk finish, vinyl acetate, 312
Silk screen ink vehicle, fluorescent, 316
Spray enamel, acrylic, white, 212-213
  polyester, white, 215-216
Spray finish, wood, vinyl acrylic, 301-302
Spray primer, 6
Stain, shingle, alkyd, red, 137-138
Steel paint, chlorinated rubber, 153-154
Stucco (and masonry) paint, latex, 160
Styrene-acrylic ceiling or wall paint, 239-240
  exterior coating, 71
  interior ceiling paint, 240
  interior flat paint, 237-239
  interior textured finish, 284-286
  maintenance paint, 364-365
  primer, anticorrosive, 6-7
  primer, maintenance, latex, 57
  primer, metal, latex, 47
  primer, wood, 36-37
Styrene-alkyd gravure ink, 318
Substrates (rigid or semirigid), latex "Hypalon" paint, 336
Swimming pool paint, acrylic-latex, 322-323
  chlorinated rubber, 158-159
  latex, 323-324
  polyvinyl chloride, 324

Temperature conversion tables, 370-372
Texture finish, exterior, vinyl acetate, 139
  interior, styrene-acrylic, 284-286
Texture paint, vinyl acetate, 286-287
Thermoplastic ink, epoxy, 317
  maleic, 317
  urea, 318-319
Thermoplastic lacquer, wood, air-dry, acrylic/butyrate, 294
Thick film-corrosion resistant epoxy coating, 363-364
Tint base, acrylic, economy, 273-274
  acrylic exterior, light, for glycol dispersed colorants, 112-114
  acrylic, interior, eggshell, 276
  acrylic, interior, semigloss, for water and glycol dispersed colorants, 272-273
  acrylic, sheen, 114-115
  Cal Ink colors, latex, 319-321
  enamel, light, latex semigloss, 177
  latex, 277-283
  latex, exterior light, 119
  latex, exterior semigloss, 115-119
  vinyl acrylic, flat, 283-284
  vinyl ethylene, exterior, 120
Tint, exterior light or white latex, 108-111
  exterior light or white vinyl ethylene, 111-112
Tintable enamel, acrylic, eggshell, 183-184
Topcoat, acrylic flexible vinyl substrate, 306
  automotive, clear acrylic, 306
  clear, acrylic urethane, 309
  exterior, acrylic white, 120-123
  exterior latex, chalking white, 125
  exterior latex, pigmented, 125-126
  exterior vinyl acrylic white, 124-125
  interior acrylic white, 241-252
  wood, catalyzed, 300

Trademark chemicals, 383-389
Trademark chemical manufacturer's, 379-383
Traffic line paint, alkyd, 330
alkyd, chlorinated rubber, 329
latex, 331-334
polyvinyl chloride, 334-335
Traffic paint, acrylic latex, 325-328
chlorinated rubber, 330-331
Triamino ester enamel, white, 211-212
Triamino ester maintenance enamel, chrome yellow, 364
Triamino ester primer, 29-31
Trim paint, exterior, acrylic, 126-136
latex, 136-137

Unsaturated polyester wood coating, 301
Urea thermoplastic ink, 318-319

Varnish, alkyd, overprint, 297
brushing, tradesales, acrylic, 298
floor, epoxy ester, 297
Ventilation, 377
Vinyl acetate exterior-interior flat paint, 240
exterior textured finish, 139
silk finish, 312
textured paint, 286-287
Vinyl acrylic board coating, satin-finish, 313
exterior emulsion paint, white, 100-102
exterior paint, 72-73
white, 100-102
exterior topcoat, white, 124-125
interior flat wall paint, 241
primer, 3-5
tint base, flat, 283-284
wood finish spray, 301-302
Vinyl automotive top-texturizer, 306-307
Vinyl chloride acrylic house paint, exterior, white, 102-104
Vinyl coating, 308-309
Vinyl ethylene exterior tint base, 120
Vinyl ethylene exterior tint, white and light, 111-112
Vinyl plastisol molding, film forming, and fabric coating, 307-308
Vinyl primer, 48-49
coil coating, 48

Wall or ceiling paint, styrene-acrylic, 239
Wall primer, interior, 5
Wash primer, polyvinyl butyral, 49
Weather barrier mastic coating, 156-157
Weldable alkyd metallically pigmented coating, 153
White baking enamels, high gloss, 343-345
water-reducible, 343-344
White epoxy enamel, solventless, thick film, 256-257
White wash, 139
Wood and masonry exterior latex paint, 159
Wood coating, alkyd modified, flexible polyester, 299-300
unsaturated polyester, 301
Wood finish, acrylic, 299
polyester, clear, force-drying, 300-301
vinyl acrylic, spray, 301-302
Wood lacquer, acrylic/butyrate, 289
thermoplastic, air-dry, 294
Wood primer, styrene-acrylic, 36-37
Wood topcoat, catalyzed, 300

Zinc chromate primer, 16–17
latex red iron oxide, 20–21
PVC, 24–25
Zinc dust primer, epoxy-polyamide, 19–20
PVC, 23–24
Zinc oxide primer, latex, 21–22
Zinc-rich primer, epoxy, 17–19